Embedded Thinking

Zsuzsanna Kondor

Embedded Thinking

Multimedia and the New Rationality

PETER LANG
Frankfurt am Main · Berlin · Bern · Bruxelles · New York · Oxford · Wien

Bibliographic Information published by the Deutsche Nationalbibliothek
The Deutsche Nationalbibliothek lists this publication in the Deutsche Nationalbibliografie; detailed bibliographic data is available in the internet at <http://www.d-nb.de>.

Cover illustration:
Zoltán Szendrő

Cover design:
Olaf Glöckler, Atelier Platen

ISBN 978-3-631-57732-5

www.peterlang.de

CONTENTS

PREFACE

The deliberations in the papers included in this volume suggest that the perspective of communications technology enables us to see traditional philosophical inquiries in a new light. More precisely, from the perspective of communications technology it becomes clear that a radical turn has become inevitable, especially in the field of metaphysics and epistemology; this perspective is a considerable help when we attempt to interpret 20th-century philosophy's criticism of its predecessors.

Characterizing the traditional disposition of philosophy as atomistic, individualistic, objectivistic, and dualistic (to mention just some of the well-known labels), is common in historical accounts. The difficulties and anomalies here arising have been thoroughly discussed, and attempts to overcome them were numerous, but success was elusive. The perspective of communications technology, of course, does not change this situation at a stroke, but it is very promising nonetheless: it furnishes the evidence that even on the most abstract level of human thinking, context in a wider sense has considerable importance. Let me recapitulate the steps which led me to this train of thought.

In the early 1920s and 30s, the Hungarian scholar István Hajnal developed a theory according to which Western rationality was intimately grounded in alphabetical writing. By the 1980s, this theory became, via Harold Innis, Marshall McLuhan, Walter J. Ong, and Elizabeth Eisenstein (to name just the main nodes in a momentous network of ideas) the standard paradigm. Amazingly, Heidegger's criticism of Western metaphysics and Hajnal's discovery of the roots of Western rationality display a spectacular convergence. Within the framework of the ongoing Budapest-based international project "Communications in the 21st Century", I tried to analyze how philosophical thinking has been changing in accordance with the modifications of communications technology. Also, I started to investigate the cognitive background of these changes. McLuhan's phrase "The medium is the message", and an important discovery by the Toronto School that different media mediate the same content differently, forced me to face the problem of widely different ways of representation. First, I compared the metaphysics of verbal and pictorial representation; here it became clear that due to the different ways of representation, different thinking habits could emerge and become dominant. Further investigations in the realm of visual perception and representation, as well as into higher cognitive processes such as categorization and the emergence of language both in phylo- and ontogenetic senses strengthened my impression that the dominant representational pattern (in Western philosophy it is written language) has an amazingly great influence on our world view.

Comparing the epistemology and metaphysics of literacy and that of the age of so-called secondary literacy, a new rationality seems to emerge. (I refer to literacy in Ong's sense, i.e. the age dominated by written records; I consider secondary literacy as an epoch which is characterized by the rationality of literacy, but allows for multimodal enhancement due to changes in communications technology.) This new rationality can best be characterized by a stress on embeddedness, the recognition of the importance of perception and action even in higher cognitive functioning, a certain distance from language as the main mediator, and (as a summing up of these elements) the emergence of the outlines of a new conceptual framework closely related to the problem of categorization. The forerunners of this imminent change are well-known philosophers of the 20th century including Bergson, James, Dewey, Wittgenstein – and Heidegger.

Traditional epistemology suggests capturing knowledge as if it were the result of mere abstract reasoning, and reasoning is regarded as if it were some computational process with clear-cut categories. The traditional dualism of mind and body, and the permanent "mental discomfort" because of the lack of a suitable language came to be seen in a different light due to the change of perspective suggested by the philosophers I have listed. The opalescent Bergsonian concept *image* considers the human being as an active element of the material world. This world is accessible due to the intimate relationship between the human body and the material world and the special active character of the human body. It is possible to accumulate the information gained and to integrate it into subsequent actions due to memory. And from the perspective of Wittgenstein's language games, language is also considered within the framework of everyday activity. Wittgenstein's ideas suggest a very broad change in the perspective of the investigation of language. Seen from this perspective, some decisive, and at the same time distorting, features of language come to light, under which the traditional framework of philosophy became questionable. The Heideggerian notion of referential totality also emphasizes the embeddedness of the human mind in the world, and considers a mostly un-reflected engagement in the world to be primordial in comparison to the analyzing attitude of Western rationality.

That is, these three philosophers propose handling practice and reasoning within a common framework, and at the same time stress the importance of context. This point of view means that human understanding and the nature of knowledge are best comprehensible not from the perspective of abstract reasoning, but rather from that of mundane practice. Recent cognitive and developmental psychological research suggests something similar when it clearly shows the importance of context on the one hand, and motor activity on the other. It is at this juncture we arrive at Lakoff and Johnson, who suggest, precisely, to consider abstract concepts as being based on perception and motor activity. A key term, *embodiment*, condenses this point of view. It means that "conceptual structure is meaningful because it is embodied, that is, it arises from and is tied to, our preconceptual bodily experiences. In short, conceptual structure exists and is understood because preconceptual structures exist and are understood."[1]

Relying on these considerations, a demand unfolds for change in critical nodes within the contemporary mainstream epistemological framework – nodes such as the notions of reality and knowledge, to name the most obvious ones. We cannot avoid using this conceptual framework but we certainly face sometimes implicit, sometimes explicit claims for reconstructing it.

This short survey clearly shows that there is a strong demand to induce a shift in philosophical thinking. Communications technology catalyzes this demand in a twofold sense: the focus on the means of communication reveals the importance of the framework of expression and sheds light on the differences caused by different media. And it enables us to see the attempts to dissolve anomalies in a different light, in that it makes it possible to comprehend these efforts as a problem of expression.

The chapters in this volume are papers I have presented at different conferences and workshops. Despite the sometimes diverging focuses of these events, the unique perspective of communications technology on the one hand, and the overlapping tendencies taking shape in recent attempts of philosophy and cognitive science on the other, in the event provided me with solid grounds to see the outlines of a theoretical convergence.

The four consecutive parts of this volume show the way I came to develop the idea that communications technology has considerable philosophical significance. There is a historical perspective running through this volume: again and again I strive to highlight the turning points and crucial insights on philosophy's way. Similarly, I attempt to present communications technology in a historical frame of reference. Since communications technology is closely bound to representational patterns, aspects of cognitive psychology and cognitive science also come into play.

In the first part, *Language and Communication*, I try to clarify the ancient notion of *logos* and its relation to language; the close-knittedness of communi-

---

[1] George Lakoff, *Women, Fire, and Dangerous Things: What Categories Reveal about the Mind* (1987), Chicago and London: The Univ. of Chicago Press, 1990, p. 267.

cation and community; the birth of modern logic and its relation to literacy; and the emerging efforts in the 20th century towards restoring the unity of human intellect and its environment. Against the background of communications technology, the way nearly identical criticisms could evolve within the framework of opposing philosophical commitments becomes apparent, an alternative to Derrida's criticism of *logo-centric* tradition emerges, and light is shed on how language can come into play as a framework of identity and commitments from a Heideggerian point of view.

In the second part, *Philosophy and Communications Technology*, the history of philosophy is examined against the background of changes from primordial orality through literacy and the printing press to electric and electronic communication. It is in the first chapter here that I focus on a rough overview of some crucial questions emerging in parallel with these changes. In the second chapter, the focus is on the modifications of the concept of reality and virtuality. In the next chapter, the concept of friendship emerges as a conspicuous element of everyday life which could not be incorporated in a coherent manner into the abstract framework of Western metaphysics. "Being Mobile: Cognitive Multiplicity" sheds light on some outworn premises of earlier epistemologies, while paving the way for the perspective of cognitive science that is discussed in the chapters following. The last chapter paves the way towards the next part of the volume by concentrating on Wittgenstein's implicit and Ivins' explicit criticisms of the literate culture/mind.

The third part then, *Images and Words*, focuses on the main differences between pictorial and verbal representations. The perhaps pedestrian point of view of mobile communications research and mobile learning is introduced, while not losing sight of the metaphysical dimension. In the first chapter of this part, beside the practical issue of MMS, multimedia messaging, there follows a comparison of Egyptian and Western metaphysics with special regard to their recording systems. In the second chapter, Heideggerian philosophy gains special importance as it emerges as a critique of the main scene of clear-cut Western rationality, namely the limits posed by verbal expression. Here I suggest that several similarities are noticeable between Egyptian thought and the thought that Heidegger believes desirable. In the next chapter, there is a short overview of MMS research, supplemented with reflection on the cognitive peculiarities of visual perception and the main differences between verbal and pictorial expressions. In the last chapter of this part, I evaluate research on mobile learning against a background of multimodality and cognitive evolution theory.

In the final part of this volume, *Language and Cognition*, I focus on the complex interplay between language and the problems of philosophy. In the first chapter, I examine, relying on Wittgenstein, some special characteristics of language. In the second chapter, with reference to certain peculiarities of verbal expression, I try to place language within the framework of cognition and at the same time say something on the misleading concept of reasoning as used in earlier philosophical traditions. It turns out that perception and action are closely related even to highly abstract reasoning. Finally, in the last chapter, on the basis of previously outlined theories as regards the cognitive background of reasoning, I attempt to point to some convergences between traditionally opposed schools of philosophy and the intriguing parallel developments coming to characterize both the humanities and the sciences.

Though having been written over a period of some nine years, the papers collected in this volume are steps, I believe, on a route constituting a single research programme. This is the programme I plan to pursue in the future, too. Insights I believe to have made during the past few years convinced me to put greater emphasis on the importance of motor activity and the productive interplay of verbal/conceptual and perceptual processing. Revisiting the Bergsonian claim, I plan to outline a frame of reference for an epistemology which is based on the "metaphysics of matter", a metaphysics which is suitable for yielding an alternative to traditional language-textured metaphysics.

As a senior research fellow at the Institute for Philosophical Research of the Hungarian Academy of Sciences, I have enjoyed considerable freedom in choosing and pursuing my research topics. Also, I am indebted to the Network of Excellence's European project Kaleidoscope, which gave me the opportunity to make the acquaintance of some wonderfully stimulating colleagues, as well as to encounter a number of challenging practical experiments and issues. These encounters helped me to see my previous publications in a new perspective, and to prepare this volume. In the course of composing it, I received numerous illuminating questions, comments, and editorial advice from Kristóf Nyíri. I am greatly indebted to him. And last but not at least, I have to thank my family for their patience and encouragement.

LANGUAGE AND COMMUNICATION

Objectification or Spiritual Involvement?*

There are some conspicuous topics in the history of communications technology which can help to reveal the connection between the media of expression and the process of thinking. These are: the birth of writing, (that of alphabetical writing and of cursive writing), the appearance of printing, the emergence of different recording systems, and the invention and use of electronic and digital tools.

In the first half of the 20th century two Hungarian authors, historian István Hajnal and classical scholar József Balogh, pointed out in an extraordinarily sensitive way that thinking and words were becoming more and more mechanized. Their work shows no similarity either in the way they unfold the problem, or in the theoretical foundations of their approach. It is all the more astonishing that similarities can be found in their respective cultural criticisms.

In this paper, I will try to find an answer to the question of how two thoughtful authors with seemingly different theoretical backgrounds can judge the present period of historical development almost identically, i.e. examine whether there is any common ground in the two theories that helps to explain the parallelism.

Hajnal started his oeuvre with paleographic research, and in the thirties he became involved in new research connected to the history of technology. He constructed a coherent theory based on two pillars: one was the consideration that methods of communication determine the development of human societies; the other was the general theory of objectification.[1]

The term *literacy*, as Hajnal understood it in the sense of *Schriftlichkeit*, is a very good example of the first pillar in connection with writing. Hajnal connects the history of writing to the influence it exerts on human thinking on one hand and social organizational forms on the other. The destiny of some very effective writing techniques developed by antiquity is a good example for the

---

* Originally published as "Objectivational Process or Spiritual Involvement?", in P. Fleissner and J. C. Nyíri (eds.), *Philosophy of Culture and the Politics of Electronic Networking. Austria and Hungary. Historical Roots and Present Developments*, Innsbruck/Wien: Studien Verlag – Budapest: Áron Kiadó, 1999, pp. 43-56.

[1] Communication as a decisive element of human societies mirrors a sociological point of view. As Hajnal pointed out, this is the living, steadily changing life contrasted with various solidified structures. Hajnal demanded this distinction – which Comte and then Durkheim had used for the first time – be conceived of as a process. The other above-mentioned pillar, i.e. the general theory of objectification, has a strong philosophical background, as Alfred Weber pointed out. According to this background, the world is divided in two parts: the reasoning subject and the object that is to be known. This tradition, which played and plays an important role in the history of philosophy, gains peculiar importance in the light of the Heideggerian criticism of metaphysics.

latter phenomenon. As Hajnal notes, the possibility of how one expresses an idea in writing depends heavily on the technique of writing. But it is also crucial for these techniques to be rooted in everyday life. They should harmonize with the needs of human interrelationships.[2]

The theory of objectification – the other pillar of Hajnal's concern – is supposed to express basically the objectification of thoughts. This is inherent in human nature which, according to Hajnal, always inclines towards some "material support". That is, human beings tend to vest their thoughts and their experience with matter-related forms. This is how they try to make ideas simpler and easier to handle, or give them a sense of timelessness. "Movements and sounds do disappear, still, humans can use them and their matter-relatedness to produce something that is objective, something that functions as an extrinsic intermediator for inner life."[3] This idea creates the foundation of the definition of technology and writing. The different instruments of technology can be seen as objectification of different kinds of knowledge regarding how to handle different raw materials. On the other hand, writing can be considered as the objectification of thoughts in written signs.

However, the act of objectification in some cases can, so to speak, appear transiently, i.e. independently from matter:

> Language picks out certain aspects of this insecure and elusive movement-sphere [i.e. human community-life], expresses them *in an artificial manner*, and transforms them into notions with a validity for all the community ... language is not solely an extrinsic instrument of communication, but it is thinking itself. ... Language creates the cognition-like preconditions for that which it expresses, creates a basis on which understanding among human beings in this form becomes possible altogether. Something that would dissolve into nothing, will be turned into ideas by the symbols of language.[4]

For any experience to become accessible for reflection, it has to be formulated, to be ordered in one or another way, so that it becomes expressible

[2] The age of antiquity achieved the level of shorthand. But the use of written records did not develop into a universal method of communication, and people remained mainly within the sphere of oral culture. Nonetheless, writing was widely used in those times and its benefits were known for those societies. "Literacy also strengthened rationalism in the peoples of antiquity...", says Hajnal (F. Glatz [ed.], *Technika, művelődés*, Budapest: MTA, 1993, p. 34), but "collective life could not yet be wholly imbued" by the use of writing. After the external framework collapsed, writing technology died away, as well. Cursive writing as a replacement to drawing individual letters was the result of a slow and long development. It is apparent that it was possible for a small intellectual stratum of society to learn writing without it developing into an effective and all-pervasive communications technology. This is how writing sank into oblivion once the institutional framework collapsed.

[3] *Ibid.*, p. 203.

[4] *Ibid.*, p. 32.

and apprehensible. Speech, a construction of immaterial sounds, is neither enduring, nor objective or extrinsic, as words are pronounced by individual persons, and personal presence is inseparable from oral communication.

When ideas are conveyed exclusively by utterance, one has to reckon with some restrictions. New ideas can only be incorporated slowly, step-by-step, into the network of traditions, and critical attitudes are also quite difficult to maintain, as attention is inevitably focused on retaining information. This is an issue of memorization and several techniques have been developed to facilitate this, such as the use of rhythm and standard patterns in all oral cultures.[5] Still, it offers less support for rational thinking procedures than writing does. Through writing, ideas become remarkably easier to handle and to elaborate, and they also become more abstract. Part of the energy previously needed for simply remembering can be liberated, and can be mobilized for other purposes. Traditions can be transmitted in another way; the role of the wise old man changes, but the organization of society can also be modified, due to the possibility of more complicated laws and legislative procedures. These are the obvious consequences of the technology of writing. However, these consequences depend strongly on the actual needs of a society and on the technique of writing. The appearance of alphabetical writing has considerable implications in this respect.

The significance of alphabetical writing has been stressed by many authors. It is the most accurate and the most abstract instrument for recording the acoustic phenomenon of speech. Havelock underscores the fact that, in respect to social control and governance, alphabetical writing was the only means to create a flexibility comparable to oral communication, because it did not ritualize and simplify the content of it: the content was not turned into any sort of authority.[6] In Hajnal's view, the most important aspect of alphabetical writing consists of the fact that it requires a certain level of abstraction. This very technological instrument made a stricter separation possible between the objectified idea and its empirical source, and the distance between practice and the world of ideas grew.[7] As cursive writing departed from the cumbersome

---

[5] On the different technologies of memorization see: F. A. Yates, *The Art of Memory*, London: Routledge and Kegan Paul, 1966.

[6] Pattison mentioned similar arguments used in Sparta to turn the Spartan administration against the use of writing. Lycurgus was afraid of writing (and thus externalizing) legislation, and making it inflexible. He felt that people will not understand jurisdiction when written down, they simply obey: they vest it with authority, disregarding its content. (R. Pattison, *On Literacy. The Politics of the Word from Homer to the Age of Rock*, Oxford: Oxford University Press, 1982, pp. 56 f.)

[7] "The alphabet, though it probably derives from pictograms, has lost all connection with things as things. It represents sound itself as a thing ..." (W. J. Ong, *Orality and Literacy. The Technologizing of the Word*, London and New York: Methuen, 1982, p. 91.)

method of drawing letters that are not joined together, it created an instrument that can be used more easily and more quickly. Thus it was suited to helping its user, after thorough training, achieve a sort of ***thinking-in-writing***.[8]

In Hajnal's view, printing appeared thanks to new social needs and tendencies. Printing was actually a result of the objectification of technology and the use of written records[9], and the result of the social needs thus engendered.10

Writing was a more artificial tool for articulating ideas than speech, and it recorded experience in a lasting form. This brought many changes. It paved the way towards the realization of some "universally human" attributes, and opened up seemingly infinite perspectives for human thinking. As the legacy of the past became accessible to subsequent generations, universal and eternal human attributes came to light, and this was a very important step in the process of human beings growing conscious of self. The relationship to the past and its legacy also changed, and new needs were formulated as sort of counterpoints to those of the past. The belief in the human mind's universal validity gained ground, and this viewpoint became more and more evident. On this basis it was natural for people to gain self-confidence,[11] in the sense of the Cartesian *cogito*, and trust in the authority of rational thinking increased. This was of course not by chance: ideas materialized, so to speak, and as a consequence, they became easier to handle, combine, and make more complex.

> Things that had earlier happened instinctively in human beings' inner and outer lives, started to take a conscious turn with the appearance of literacy. This sphere of life becomes *objectified and abstracted*; the human being projects this sphere in front of

---

[8] In Hajnal's estimation, cursive writing appeared first in the 13th century, i.e. in the epoch of "magisterial" writings. "Cursive writing was felt at that time as a new invention, people were not yet accustomed to freely using it. They learnt it slowly and cumbersomely, and insisted on certain rules." (I. Hajnal, *Írástörténet az írásbeliség felújulása korából*, Budapest: Budavári Tudományos Társaság, 1921, p. 23.) Cursive writing was still a masterly performance – in the double sense that it came close to being a work of art, and that it was practiced, decisively, by the graduates – the "magistri" – of the University of Paris.

[9] "The printing of books was, as is well-known, the product of the exact melding of many details of the magisterial transcription of work, of school dictata, of paper manufacturing, of several types of inks/paints, of the metal industry and the finest engraving craft." (***Technika, művelődés***, p. 228.)

[10] "The invention of the printing press was only an inevitable consequence of the deep-rooted and large-scale writing culture of the late Middle Ages. As many people understood written texts, there was a large reading audience and also wide-scale book production in the last centuries of the Middle Ages, which brought forth by pure practical expediency the invention of printing." (***Technika, művelődés***, p. 24.)

[11]See Heidegger on the self-confidence of the subject in modern history; and the dysfunctionality of doubt that is irreconcilable with orality. Cf. K. Nyíri, "Tradition and Communication", in: G. Jaritz (ed.), *The Art of Communication*, Graz: Akademische Druck- u. Verlagsanstalt, 1995, pp. 317 f.

himself, and examines it *consciously and from the outside*. There arises the possibility for methodical purposefulness, for the conscious handling of concepts, and for combinational and complicated work.[12]

However, these benefits are treacherous, because objectification creates a world in which all phenomena become detached from their origins and obey some other kinds of rules than before. The original cause fades away, but the ideas it generated start out into a new, independent life in their written form. They become objects of criticism, and they, themselves, will generate new ideas as well.

Hajnal's critical attitude, whether he is speaking about technology, writing activities, or so-called intellectual work, always concerns over-mechanization, which is inseparable from the process of objectification. The same one-sidedness, the same inability to experience creativity, the same empty uniformity is characteristic for all three fields. What is the reason for this? As Hajnal points out, writing has its benefits in enhancing people's self-confidence, in facilitating and speeding up the process of thinking, and in organizing societies. But all these benefits are consequences of one single quality of writing: that it has an objectifying nature. Hajnal speaks about the over-technologization of the "conceptual construction of thinking", and underlines that a very long period of development was needed for writing to become flexible enough to be able to very narrowly follow the way of "thinking-in-utterances". Writing is actually an artificial instrument for thinking, one through which ideas can free themselves from the context of real life.

Letters produced letters, writing produced writing. Naked and speculative thinking used to be very refined, even in the smallest detail-exercises. The more one-sidedly it functioned, the more it sliced professions into mechanical details, the more it exempted them from sensing life's heavy material, the more perfect it was.[13]

Hajnal very clearly describes that the process of human activities becoming empty and over-mechanized is rooted in the nature of objectification. That is: everything that is recorded or fixed, takes up a material form, starts a new life of its own, and the laws of its further functioning will be independent from its

---

[12] *Technika, művelődés*, p. 18.

[13] *Ibid.*, p. 449. Hajnal thinks it is determinative that mental and physical work should belong together, and they should interact closely. No happiness can be found in any activity restricted to the material, or restricted to the world of letters, he claims. This is true, considering the picture he has about human beings. But it also generally holds under the conditions of inevitable objectification. A separation from a certain sphere that is ruled by certain other rules makes it possible that the mechanism determining a given sphere will overrule everything else. However, Western development was the result of long and recollected work, and the constantly refined specialization of it. This specialization split up each form of activity, the result of which was one-sided and comfortless mental and physical work.

origins. This is what happens in the case of writing: the body of texts is growing, but this process is prompted more and more by a certain pressure deriving from the texts themselves, and not by everyday life.

József Balogh kept searching for evidence proving the existence of the practice of reading and writing aloud. The sections quoted by him show that living speech has not only aesthetic values, but its relation to the intellect is also direct and of primary importance. Thus, there can be no real understanding without the simultaneous functioning of visual and acoustic experience. These texts primarily show a commitment to the view that living speech has a much greater convincing force than written words. It turns out that "content and intellectual joys, intellect and pathos were one single and indivisible unity" for the ancients.[14] Style and perfection of form was, of course, much more important for a writer in those ages than for anyone in the period of silent reading. The activities of *audire*, *legere*, and *intelligere* constituted one single unity.

In light of the entire direct and indirect body of evidence, the following question arises: why did reading become silent? The first reason was printing. The invention of printing started a process that Balogh calls the over-mechanization of the word. As printed texts were easier to read, reading became quicker. Complicated abbreviations ceased to be used, and words were clearly recognizable. Balogh regards all instruments of communication, (such as the typewriter, dictation and speaking machines, the telegraph, and the telephone), as tools of mechanization. These instruments made reading and writing more comfortable, but this convenience resulted in superficiality and speediness through which the "art of writing degenerated into a kind of factory production".[15]

As we can see, Balogh's deliberations differ basically from those of Hajnal. On a textual level, this difference becomes evident in the question of printing. Hajnal stresses the social element of literacy and technology. In his view, printing was a fortunate junction of these dimensions. Balogh examines, in turn, the long-term consequences of printing and does not touch upon the issue of the technical invention itself.

The not-explicitly-formulated theoretical base of Balogh's considerations can be easily reconstructed with the help of the texts he quotes, which show a certain similarity to Heidegger's views. Heidegger says that in the course of the history of metaphysics, in the epoch from Anaximander to Nietzsche, the truth of *Being* remained hidden.[16] The reason for this was the fact that the power and authentic role of language was not recognized, although this is a crucial motif in

---

[14] J. Balogh, *Voces Paginarium: Adalékok a hangos olvasás és írás kérdéséhez*, Budapest: Franklin Társulat, 1921, p. 10.

[15] *Ibid.*, p. 29.

[16] M. Heidegger, *Wegmarken* (*GA9*), Frankfurt a. M.: Vittorio Klostermann, 1976, p. 369.

the problem of *Being*. Heidegger intended to revive and reformulate this forgotten tradition.

Balogh quotes from Nietzsche, whose meaning was either not clear, or misunderstood, or forgotten in the latter's time. These passages emphasize the magnificence of ancient poesy as opposed to the pale, abstract, and inanimate writings of Nietzsche's time. Balogh mentioned it, of course, in the context of reading aloud, where it should also be noted that it makes a huge difference whether a piece of writing is meant for loud or for silent reading. The belief in the power of the uttered word goes back to the interpretation of *logos* in early antiquity. *Logos* has a holistic meaning, and this is partly why it is difficult to translate. We have found the following interpretations according to the enumeration of Heidegger[17] language, word, speech, reason, intellect, explanation, proportion, necessary structure. These translations show that it can be defined as a sort of force or ruling principle, but also as an articulation thereof. Anaximander touches upon this point, when out of the infinite, the limitlessness (*aperion*) – which he considers as the *arkhe* – arose the ability to define the borders of things and entities through *logos*, and as a result of this, chaos will be replaced by order.

This interpretation of logos involves an intimate relation between the pronounced word (which is inseparable from the person uttering it), thought, and language.[18] In the light of this deliberation, we can understand what Heidegger intended to state when he wrote

> Language is the house of Being. In its home man dwells. Those who think and those who create with words are the guardians of this home. ... Thinking acts insofar as it thinks. Such action is presumably the simplest and at the same time the highest, because it concerns the relation of Being to man.[19]

---

[17] M. Heidegger, *Sein und Zeit*, Halle a. d. S.: Max Niemayer Verlag, 1927, p. 32. I used the transl. by M. Heim, *Electric Language: A Philosophical Study of Word Processing*, New Haven and London: Yale University Press, 1987, p. 34.

[18] Cf. "Heraclitus suggests that language deepens thought in its operation; it does not remove the thinker from what is being said. Far from being a tool of thought, language, or logos is the element of thought. Heraclitus insists: What we think about language is implicated in how we think in language; and alternatively, the manner in which we think in and through language entails a certain stage or degree in our understanding of language." (Heim, *Electric Language*, p. 37.)

[19] Martin Heidegger, "Letter on Humanism", in: David Farell Krell (ed.), *Basic Writings – Martin Heidegger* (revised and expanded edition), London and New York: Routledge, 2007, p. 217 (first published in 1947, Bern: A. Franke Verlag). As R. E. Palmer recapitulated this famous topic: "Ontologically stated, what a thing *is*, it is through the openness of language. Being is inseparable from the operation by which language causes earth to shine forth 'in a certain light'. ... being, as the process by which language brings things to the open, is linguistic in its nature." (Palmer, R. E., "The Postmodernity of Heidegger", in: W. V. Spanos,

Heidegger says that for the Greeks, Being only speaks out when Being comes out of hiding. Being happens originally (anfänglich) from the word, because word substantially means that it lets unhidden Being appear and come to light, and be preserved. Speech names something, but this naming is not a simple labeling, it is evocation. And this evocation shows the profound relationship between things and the world, thus it shows the difference between things and the world. This separation from-yet-among one another, this Unter-Schied is neither a distinction, nor a relation between things and the world. It is a context, one in which things and the world can be grasped in their cwn specificity without distortion. Language exists as the separation from-yet-among one another manifesting for the world and the thing.[20]

According to Heidegger, language can bring to presence the intimate interrelationship of human beings, Being, and the surrounding world. In order to see this, one has to be conscious of this possibility, that is, language must not be used simply as a tool for transmitting information, but it should be regarded as a medium possessing an elementary revelatory force, in which the human being is essentially determined to move.

Heidegger also emphasizes the problem of writing. He does not stress the difference between utterance and writing[21], at least not in connection with spiritual involvement. He refrains from doing so because he finds that human hands can represent the same immediateness in relation to Being as utterance does. Writing is in the "showing" that happens through the hand, appearing as word. The original form of writing is handwriting.[22] Printing brings a fundamental change: instead of the word's picture (Wortzeichen) there comes the letter – set and printed.[23] The printing press and typewriter made writing easier and more effective, but, at the same time, it covered up its original

---

(ed.), *Martin Heidegger and the Question of Literature. Toward a Postmodern Literary Hermeneutics*, Bloomington: Indiana University Press 1979, p. 83.) About the interrelatedness of saying, Being, word, and thing, see M. Heidegger, "Das Wort", in: *Unterwegs zur Sprache*, Tübingen: Günter Neske, 1959, pp. 217-238.

[20] See Heidegger, M., "Die Sprache", in: M. Heidegger, *Unterwegs zur Sprache*, p. 30.

[21] This is a disputable point of view as regards the difference of orality and literacy. "Writing is in a way the most drastic of the three technologies. It initiated what print and computers only continue, the reduction of dynamic sound to quiescent space, the separation of the word from the living present, where alone spoken words can exist." (Ong, *Orality and Literacy*, p. 82.)

[22] "The word indicated by the hand and appearing in such marking is writing… Being, word, gathering, writing denote an original essential nexus, to which the indicating-writing hand belongs. In handwriting the relation of Being to man, namely the word, is inscribed in beings themselves." (Martin Heidegger, *Parmenides* (1982), trans. by A. Schuwer and R. Rojcewicz, Bloomington and Indianapolis: Indiana University Press, 1998, pp. 84 f.)

[23] "This mechanism of setting and pressing and 'printing' is the preliminary form of the typewriter." (*Ibid.*, p. 85.)

authenticity. What is more, this happened unnoticed. A new mechanism was brought about, which became part of everyday life, and which fundamentally changed the relationship of hand and word. Typewriting is hardly anything more than a mechanical movement of the hand. This change (i.e. the connection between hand and word) – Heidegger claimed – altered our relationship to the world and to things as well.

The human hand plays a very important role in writing. Balogh notes that in over the course of time, transcribing and writing in old monasteries became less and less audible. The reflection upon this phenomenon stressed the importance on the threefold-unity of the mind, the eye, and the hand.

It is language, says Heidegger, that produces the conditions according to which world and all existing things can demonstrate themselves in their separateness to man (cf. Anaximander). This can happen through utterance and through handwriting also, but only when there is a direct relationship between human beings and language without distortion. If a tool distorting the relationship of language and man is used, the relation itself will also become distorted. The revelatory force of language will be weakened if mechanisms determining the tools of expression appear. And according to Heidegger, this is exactly what happened in the case of typewriting.

Balogh regards printing as the starting point of the regrettable process of the word's mechanization. In the long run, he says, there is an inevitable devaluation of human utterance, and writing becomes a factory profession. The over-mechanization of reading and writing can be considered a distortion of activities closely related to intellectual effort. In this respect, Balogh's considerations are very similar to those of Heidegger.

Hajnal states that he is devoted to materialism, a devotion which derives from accepting the concept of objectification. József Balogh, on the other hand, formulates a criticism that falls in line with a tradition that stresses the importance of the spirit, or even the logos-like power of language and intellect. Still, he grasps the problems of his epoch in a very similar way to Hajnal: writing and reading became a mechanic activity, and as such, it inevitably became estranged from life. Part of my initial question has an answer now, since a viewpoint stating the identity of language and thought was expressed, and also, the importance of spiritual involvement – according to the tradition of logos – has entered the picture.

As Hajnal says, language is thought itself, that is, language makes it possible to transform our experiences into thought-content. Otherwise, they would continue to be an obscure whirl of sensations. Without the structure-forming force of language, it is impossible to speak about reasoning. In the tradition of *logos*, *logos* is the element of cognition. This means, using Heim's terminology, that language affects cognition as far as the horizon where thoughts appear. These thoughts are formed by language. In this respect, language is not simply a medium for articulating thoughts, but it also plays an important role in constructing the thought that is going to be formulated.

The medium of the articulation of thoughts can be regarded – in the tradition of logos – as an element affecting cognition. All the changes that occurred at the introduction of printing can be considered as some matter of fact proof of this. Beginning in the ancient ages, through the Middle Ages, the theory of *logos* was preeminent, even if a slight modification can be traced in its interpretation. One of the effects of this doctrine was that printing has long been regarded as something undesirable, because of the aforementioned lack of personal spiritual involvement. The long upheld tradition of reading aloud is also relevant evidence, and the fading away of this habit was a consequence of printing. After the emergence and spread of printing, the conception of cognition and language altered. Method became dominant in the construction of thinking, and the problem of language became manifest primarily as a logical or grammatical question.[24]

One quite important factor behind these changes was the fact that printing as a technological invention had a much stronger presence than any other invention before it in the sphere of writing. This is why the "taken-for-granted" character[25] of this technology could be more effective. And its efficiency and convenience had more weight than any argument on the level of principles brought up against it. In the light of Hajnal's objectification theory, we can state, as Balogh did, that printing and its consequences pointed in the direction of mechanization.

Through the use of printing, a new and extrinsic tool is wedged in between thought and written text, and this tool becomes totally independent in its functions from human intellect, which strives to be either basically open towards the transcendence of *logos* or needs a closer relation to everyday life.

Both Hajnal's and Balogh's theories can serve as a good source for criticism of modern communications technology: digitalized signs are subordinated to certain mechanisms, just as printing, and they are strange to real

---

[24] Writing as a way to make our thoughts objective and then the printing press which made the resulted changes even more radical are necessarily conducive to the idea that language is not a transcendent power able to order our world, but rather an instrument which is to be improved for the sake of higher efficiency.

[25] "Technological cultures constantly introducing useful materials and objects into the world. Because they are useful, they are accepted without thought. Sometimes they are so retiring that they are hardly noticed unless they are being used. Their relative invisibility has very little relation, however, to their influence on the shape of consciousness." (O. B. Hardison, *Disappearing Through the Skylight. Culture and Technology in the Twentieth Century*, New York: Penguin, 1989, p. 83.) To change the perspective: "A truly new technology refuses to stay classic. Even if it was first created for a classic function, it eventually becomes expressive and reshapes the function." (*Ibid.* p. 237.)

life, just as the world of printed signs. However, communications technology, having until recently engendered a process of alienation, now seems to support a rather natural environment for everyday activities. [26]

As we can see, Hajnal's criticism is based on the theory of objectification. Balogh's criticism has its foundation in the reflections of a scholar, and its theoretical background has roots in the philosophical tradition of the conviction that language is more than a mere means of transmission. Heidegger provides elaborated descriptions of and arguments for this tradition. Heidegger was explicitly opposed to Cartesian dualism, and his main endeavour was to immerse the reasoning subject into the world of mundane practice. This demand is conspicuous if we consider some key terms from *Sein und Zeit* and, of course, his criticism concerning the typewriter. That is, mundane everyday activity has a far-reaching effect on the comprehension of reality and on metaphysical insights. On the other hand, although Hajnal's theoretical background apparently embraced the Cartesian divide between the reasoning subject and his/her material environment (res cogitans and res extensa), his main concern was to show the unity of these elements.[27]

Summing up the insights from our comparison of Hajnal's and Heidegger's (and thus Balogh's) criticisms, it becomes apparent that though their metaphysical backgrounds and the main concerns are different, the recognition of the importance of everyday routine and the effort to include it in the framework of metaphysics on one hand, and philosophy of history, and paleography, on the other hand, creates a background for potentially converging conclusions.

[26] Cf. Nyíri, K., "The Mobile Telephone as a Return to Unalienated Communication", in: P. Glotz, S. Bertschi, and C. Locke (eds.), *Thumb Culture: The Meaning of Mobile Phones for Society*, Bielefeld: transcript, 2005, pp. 161-170.

[27] His view on the human nature, technology, and even history tries to catch the manifestation/crystallization of ideas, certain skills, and customs in everyday life.

The Tradition of *Logos* and a New Intellectual Attitude*

In the following chapter, I would like to consider whether the nowadays with fervour criticized tradition of *logos*[1] can help us find a new kind of mental disposition, a new horizon of concepts to grasp the world. In doing so, I will rely mainly on the Heideggerian concept of language on the one hand, and the results of the Toronto Circle on the other, especially on those of Walter J. Ong and Eric A. Havelock.

Western thinking is determined by a conceptuality which originates from the tradition of *logos*. Interpreting what *logos* meant in ancient Greek philosophy, or even to Heraclitus, is of course notoriously difficult. According to the fragments of Heraclitus, *logos* can be comprehended as some sort of power or ruling principle, and at the same time the expression of it.[2] Heidegger suggests that "the basic signification of λογοσ is 'discourse'" and at the same time emphasizes that it

> gets constantly covered up by the latter history of the word the λογοσ, and especially by the numerous and arbitrary Interpretations which subsequently philosophy provided. Λογοσ gets 'translated' (and this means that it is always getting interpreted) as 'reason', 'judgment', 'concept', 'definition', 'ground', or 'relationship'. But how can 'discourse' be so susceptible of modification that λογοσ can signify all the things we have listed, and in good scholarly usage? 3

In accordance with my hypothesis, these modifications are connected to the means of expression of thought, i.e. communications technology. After the invention and spread of alphabetical writing, an intellectual disposition evolved which required rational, systematic, and logic-ruled reasoning that must be precisely verbalizable. *Logos* became a mere instrument to describe the world.[4]

---

* First published in Jeff Bernard and Gloria Withalm (eds.), *Semiotische Berichte – Ikonische Zeichen. Auswahlakten zweier Konferenzen*, Wien: ÖGS 2000/1-4, pp. 147-159.

[1] I have Derrida's criticism of *logocentrism* in mind (Derrida, *De la grammatologie*, Paris: Minuit, 1967). However, I believe that the right direction to find a way out is to follow the Heideggerian criticism of Western metaphysics, i.e. to try to step out of the immanence of the human mind.

[2] See the fragments of Heraclitus: 1, 2, and 50, in: G. S. Kirk, J. E. Raven, and M. Schofield, *The Presocratic Philosophers: A Critical History with a Selection of Texts*, Cambridge University Press, 1957. See furthermore M. Heim, *Electric Language: A Philosophical Study of Word Processing*, New Haven and London: Yale University Press, 1987, pp. 30-40.

[3] Martin Heidegger, *Being and Time*, p. 55.

[4] "In the course of Western history, *logos* changes from the event of the manifestation of beings to an instrument by which man gains control over the forces of nature." (Zimmerman,

By the 20th century, these requirements became problematic. Knowledge that is hardly, if at all, expressible by propositions gained importance. Because it is very difficult to reconcile the requirements of a clear-cut rationality expressed by propositions and those of non-propositional impressions and skills, a demand for a new horizon of concepts and a new comprehension of reality emerged. To find this horizon, fine arts and particularly poetry seem to be convenient mediators.

The connection between communications technology and the alterations of the concept of *logos* can be proved by a parallel survey of the history of communications technology and that of philosophy. Accordingly, first I will try to describe the special features of a primary oral culture, and then those of the age of literacy.

According to Ong's terminology, a primary oral culture does not possess any knowledge of writing.[5] This means that there is no way to preserve events and ideas of the past outside of the human mind and outside the domain of live communication. This fact implies some restrictions. The burden of memorization makes it difficult to incorporate new elements in the network of knowledge, although several techniques evolved to help memorization.[6]

Because of the oral manner of passing information to the next generation, a special language evolved that is – in Havelock's terms – a storage language.[7] This kind of language is a regulative one, i.e. it determines attitudes, relatedness to each other, and beliefs.

This way of expression bears some special features. It is not subordinative, but rather additive. It is redundant, which helps in following the story line. It is aggregative rather than analytic, which is in harmony with thinking in formulas. It is conservative and traditionalist, i.e. the new elements of a story could never entirely displace the old ones. Furthermore, problems and concepts are in very close relation with everyday life and things: i.e. abstract and neutral concepts are absent from this language. It is agonistic in its manner of verbalization, constructing a narrative, which will aid the speaker, or rather the singer, the *rhapsodos*, to readily preserve knowledge. Yet at the same time it works as an intellectual challenge due to its close relation to everyday life. It is empathetic

---

M. E., "Beyond Humanism: Heidegger's Understanding of Technology", in: Sheenan, Th. (ed.), *Heidegger. The Man and the Thinker*, Chicago: Precedent Publ. Inc., 1981, p. 220.)

[5] Walter J. Ong, *Orality and Literacy: The Technologizing of the Word*, London and New York: Methuen, 1982, pp. 1 and 11.

[6] On the different technologies of memorization see: F. A. Yates, *The Art of Memory*, London: Routledge and Kegan Paul, 1966.

[7] "The language of Homer is a storage language devised orally for the purpose of survival." (E. A. Havelock, *The Muse Learns to Write. Reflections on Orality and Literacy from Antiquity to the Present*, New Haven and London: Yale University Press, 1986, p. 59.)

and participatory, which entails a communal commitment to the known. Objectively-distanced knowledge was an unimaginable concept at the time. It is homeostatic, i.e. the integrity of the past is determined by the integrity of the present; and the words used are situational rather than abstract, i.e. the meaning of a word is subordinated to the situation.[8]

The features of orally constructed expression roughly outlined above show a society in which members are involved in their everyday life so deeply that a critical attitude to tradition and a distanced analysis of the transmitted knowledge are very difficult. Questions of originality, creativity, and personality are strange and indefinable in this *noetic*[9] world.

Such questions emerged after the invention of alphabetical writing and the printing press. Due to this change in the manner of the expression of thought (i.e. a change in communications technology), several changes resulted in the *noetic* world. Thanks to writing, it was much easier to incorporate new experiences and information into the network of previous knowledge. The possibility of combining ideas opened up, and at the same time social life also changed due to the possibility of elaborate legislation and, of course, due to the possibility of recording the knowledge of the present and that of the past.

According to Ong's theory, the use of writing brought an important change in the *sensorium*. "By the sensorium we mean here", writes Ong, "the entire sensory apparatus as an operational complex."[10] The importance of the invention of alphabetical writing with respect to the sensorium rests in the transformation of the acoustic world of speech into a spatial, visual world.

> Sight isolates, sound incorporates. Whereas sight situates the observer outside what he views, at a distance, sound pours into the hearer. Vision dissects, as Merleau-Ponty has observed. Vision comes to a human being from one direction at a time: to look at a room or a landscape, I must move my eyes around from one part to another. When I hear, however, I gather sound simultaneously from every direction at once: I am at the center of my auditory world which envelops me, establishing me at a kind of core of sensation and existence... You can immerse yourself in hearing, in sound. There is no way to immerse yourself similarly in sight.
> By contrast with vision, the dissecting sense, sound is thus a unifying sense. A typical visual ideal is clarity and distinctness, as taking apart... The auditory ideal, by contrast, is harmony, a putting together.[11]

---

[8] See W. J. Ong, *Orality and Literacy*, pp. 36-46.

[9] The etymological root of this word is the Greek νoῦς or νόος which means mind or intellect. It also means "common sense".

[10] W. J. Ong, *The Presence of the Word. Some Prolegomena for Cultural and Religious History* (1967), Minneapolis: University of Minnesota Press, 1981, p. 6.

[11] Ong, *Orality and Literacy*, p. 72.

I do not entirely share Ong's above-quoted thoughts, since I think this distinction of hearing and sight mirrors a literate mind.[12] It echoes the difference between the impression of taking part in live intercourse and of reading a text. As I am trying to show, mental disposition and the framework of reference of thinking have changed since the invention of writing. These changes have had considerable impact on demand as regards the ideas to be formulated and, consequently, on vocabulary as well.[13]

Because of the ability to record ideas, it became possible to share and pass on new thoughts without having an impact on either the preservation of knowledge from the past or that of the present. Language also changed, since numerous technical tools (formulas, the verse form) became superfluous. Technically speaking, there were fewer obstacles to substantial, accurate expression.

In Hajnal's view, to mention a Hungarian historian who started his life's work with paleographic research and achieved far-reaching results, the most important aspect of alphabetical writing consists in the fact that it requires a certain level of abstraction. This very technical instrument made a stricter separation possible between the objectified idea and its empirical source, which created the basis for a growing gap between practice and reasoning.[14]

It takes quite a long time to get from the invention of alphabetical writing to the ability of *thinking-in-writing*, to use Hajnal's phrase. There are several distinguishable steps,[15] and several well-formed counterarguments arose against writing. Some of the more noteworthy critics were Lycurgus,[16] who was afraid of writing as it externalized legislation; Plato, who argued against writing as an

---

[12] The distinction of vision as a dissecting sense and hearing as a unifying sense assumes that the outer world is opposed to the human being, i.e. it mirrors the presuppositions of the dualism of object and subject. Recall the effort of Heidegger to break with this tradition in his *Being and Time*.

[13] On the emergence of general notions (goodness, beauty, intelligence, courage, temperance, justice, etc.) see E. A. Havelock, *The Greek Concept of Justice. From Its Shadow in Homer to Its Substance in Plato*, Cambridge and London: Harvard University Press, 1978.

[14] Cf. Ferenc Glatz (ed.), *Technika, művelődés*, Budapest: MTA, p. 18. (See above, pp. 6 f.)

[15] Hajnal distinguished between three periods of mediaeval writing: (1) the crisis of writing, which endured till the 10-11th centuries. In this period the technology and the usage of writing declined; (2) the epoch of magisterial writing, which reached its culminating point in the 13th century. During this time, a writing technique evolved which required a long and cumbersome exercise but it was nonetheless suitable to become the background for (3) cursive writing. The importance of this distinction is the interrelatedness of the attitude towards writing and of the technical circumstances, i.e. the technique of writing itself. It is important to stress that there is a considerable difference between the cumbersome method of drawing letters not joined together and cursive writing.

[16] R. Pattison, *On Literacy. The Politics of the Word from Homer to the Age of Rock*, Oxford: Oxford University Press, 1982, p. 56.

instrument which destroys the memory and weakens the mind;[17] and Thomas Aquinas, who wrote that "Only hearing can fully be believed".[18] But all these authors wrote, and cursive writing became an instrument through which thinkers and poets could easily and obviously articulate their thoughts. The practical usefulness and efficiency of writing was beyond question. As I indicated before, writing is the necessary condition for comprehensive, reflexive, abstract, and analyzing thought.

The internalization of writing, and later, the printed word, created the appropriate conditions for silent writing and reading; intellectual activity gradually became individualized, isolated from other people and detached from other kinds of activities. Additionally, since the heritage of the past also became accessible through individual effort, reasoning and accumulated knowledge seemed to be an individual achievement. As the legacy of the past became accessible to subsequent generations, certain universal and eternal human characteristics came to light. The relationship to the past and its legacy changed as well, and new needs were formulated as sort of counterpoints to those of the past. The belief in the human mind's universal validity gained ground, and this viewpoint became more and more evident. These capabilities and the concomitant conditions gave birth to so-called Western rationality.

This Western rationality, however, is burdened with a considerable disadvantage of the written manifestation of knowledge.

> What is written leaves him, and what is written is never an adequate realization of that part of the world it discusses. Man the writer sees his life ebbing away in a stream of imperfect actualizations. Each piece of writing reinforces the feeling that the language is no longer a fluid connection with reality but instead a kind of imperfect eulogy for thoughts, events and passions that never captures the wholeness of the thing written about.[19]

The printing press strengthened these features of writing. Inter-textuality vanishes: the printed book inspires the reader to see it as a self-contained whole. The comprehensive and reflexive manner of cognition gained greater ground due to the emergence and accessibility of scientific texts. The printing press provoked nearly the same counter-arguments as did writing itself earlier.[20] After

---

[17] See Plato, *Phaedrus* 278a and Plato, *VII. Letter* 341d.

[18] "Sight, taste, touch are deceived in regard to you.
I believe whatever the son of God has said,
There is nothing truer than;
the Word of Truth." (Pattison, *On Literacy*, p. 78)

[19] Pattison, R., *On Literacy*, p. 59. Cf. Heim, *Electric Language*, p. 37 (see above, p. 9).

[20] Hieronimo Squarciafico argued against printing books with arguments similar to those with which Plato argued against writing. See Ong, W. J., *Orality and Literacy*, p. 80. But again, the usefulness and efficiency of the printing press is beyond question. It led to science developing

the invention of printed books, the concept of author, originality, and creativity were born. This is the result of the end of the era of intertextuality. In the age of the printing press, the old Chinese maxim that "a picture is worth a thousand words" gained wide acceptance.[21]

> Leibniz considered the Chinese characters the prototype for an ideal philosophical system of notation, free of the ambiguities of our spoken language and directly correlated to the objects of thought without the distorting intermediacy of spoken language.[22]

If we recall the ancient meaning of *logos*, we may note that by the 16th century, language (earlier seen as having a creative power) became more or less an instrument that should be improved for the sake of exactness. In the 16th and 17th centuries, the usage of language, as well as the appropriate method of reasoning (I have in mind the Cartesian method), was an important topic in respect of gaining exact knowledge. *Logos* as a ruling potential had been forgotten and thinking was defined by Hobbes as a kind of calculation. And, as can be seen in the case of Leibniz, speaking became a distorting and ambiguous realm of language. These changes can be considered as reflections of the change in communications technology (think of the ease and the limits of recording by mute words) and the consequences of *logos* being converted into logic.

Leibniz's proposal, i.e. that Chinese writing would be a convenient prototype for a correct language for philosophy, mirrors a quite ambivalent situation. Alphabetical writing implies the separation of the signs and the represented thing. This abstraction helps to strengthen reflective and analytical thinking, and made an abstract and elaborated conceptual framework for such thinking possible. However, the language thus evolved became problematic by the 17th century, as it was not transparent and unambiguous enough. Chinese writing, being a form of ideographic writing, shows some relation between characters and the represented things or concepts,[23] and perhaps this feature of Chinese writing prompted Leibniz to think it less ambiguous than alphabetical writing, which records the uttered words sound-by-sound. However, the demand for a

---

at a never-before experienced rapidity. Elisabeth Eisenstein explains this phenomenon, in part, as the product of the ability to precisely reproduce maps and different illustrations. (See Elisabeth Eisenstein, *The Printing Press as an Agent of Change: Communications and Cultural Transformations in Early-Modern Europe*, Vols. I-II, Cambridge: Cambridge University Press, 1979.)

[21] *Ibid.*, vol. I. p. 69.

[22] Pattison, *On Literacy*, p. 34.

[23] As Ong puts: "One kind [of pictographs] is the ideograph, in which the meaning is a concept not directly represented by the picture but established by code: for example, in the Chinese pictograph a stylized picture of two trees does not represent the words 'two trees' but the word 'woods'" (Ong, *Orality and Literacy*, pp. 86 f.).

clear, transparent language is the consequence of highly abstract and analytical thinking.

By the 18th century, trust in reason had not only become triumphal but also showed signs of wavering; and the idea of a less calculable and less clear-cut sphere of human life began to gain importance. The personal sphere as opposed to the universally human, emerged as a positive force in connection with the concept of culture.[24] Thus, the triumph of reason was followed by the questioning of its omnipotence. This tendency became stronger in the past 150 years: Nietzsche emphasized the magnificence of ancient poetry against the pale, abstract and inanimate writings of his time;[25] a question formed by the writer Tolstoy inspired Max Weber to delimit the borders of science;[26] and Heidegger revived the *logos* doctrine, just to mention some well-known authors of the epoch.

In the age of secondary orality, where orality is based on literacy and carried by various electronic instruments such as telephones, radio, and television,[27] Western rationality seemed to achieve a culminating point. But at the same time the possibilities opened up by certain technical instruments blurred the validity of an array of customary values. The huge volume of information on different cultures and the theoretically elaborated hermeneutic attitude make normative statements relative, and computer networking as a quick and user-friendly invention of communications technology raises the question whether speaking about authorship and originality still has meaning.

From the age when cursive handwriting became widespread and then printing was invented, one notices an increasing trust in reason, as well as a dominance of the perspective of the human subject as contrasted with the external world, and on the other hand an insistence on method. In the age of computers and networks, these constituents have changed: the enormous quantity of information urges the user to know more and more details and this results in a feeling of incompleteness; the possibilities provided by computers help return us to orality insofar as more informal, illustrated, and figurative means are used which make communication faster and more simple due to less

[24] If we examine the change in the concept of culture, we notice that the role of reason is surpassed by other human attributes. Cf. Pestalozzi's notion about the difference between culture and civilization, or Schiller's theory about aesthetical culture. For the concept of culture in German philosophy, see Kopp, B., *Beiträge zur Kulturphilosophie der Deutschen Klassik. Eine Untersuchung im Zusammenhang mit dem Bedeutungswandel des Wortes Kultur*, Meisenheim am Glan: Verlag Anton Hain, 1974.

[25] See J. Balogh, *Voces Paginarium: Adalékok a hangos olvasás és írás kérdéséhez*, Budapest: Franklin Társulat, 1921, p. 27.

[26] See Max Weber, "Wissenschaft als Beruf", in: *Gesamtausgabe*, vol. I/17, Tübingen: Mohr, 1917.

[27] See Ong, *Orality and Literacy*, p. 3.

complicated expression and to the easier understanding resulting from the former. This is not equivalent to personal presence, but it recalls it, inasmuch as understanding can rely not only on verbal but even on pictorial expression. Accordingly, some of the ideals of literate culture came under question.

The thorough change of life has been reported for a long time. It has been observed, however, that trust in the fine arts as a key for or against this change had increased during the last century. We have seen that the importance of the individual person had already commenced to increase in the 18th century. In the 20th century, fine arts became a sphere from which a new horizon of concepts was hoped to have emerged. Finding another kind of conceptual framework seemed to be necessary, because the concepts used, to refer to the widely voiced diagnosis, were already unable to capture our world. This statement gained ground from two different points of view, one of which

> examines five basic and interrelated areas – nature, history, language, art and human evolution – reviewing the ways in which central concepts in each area have changed since the beginning of the present [20th] century. Because the changes have been fundamental, the concepts – and even the vocabularies and images in which the concepts tend to be framed – no longer seem to objectify a real world. It is as though progress were making the real world invisible.[28]

Heidegger provides the other perspective from which the present horizon of concepts is found inadequate In accordance with the ancient concept of *logos*, Heidegger considers language a revelatory force closely related to human beings and to *Being* itself.[29] Heidegger claims that it is thinkers and poets who are capable of preserving the force of language. To "preserve *the force of the most elementary words* in which Dasein expresses itself" is the ultimate task of

---

[28] O. B. Hardison, *Disappearing Through the Skylight. Culture and Technology in the Twentieth Century*, New York: Viking, 1989, p. xi.

[29] According to Rorty's interpretation, this interrelatedness of language, human beings, and Being, should represent some kind of warning against the old metaphysical dualism of subject and object. The distinctions between these elements "are being deliberately and systematically blurred. I read this blurring as a warning, analogous to Wittgenstein's, against trying to get between language and its object, plus a further warning against trying to get between language and its user. In Heidegger's version, the warning says: if you try to come between them ... you risk winding up with some version of the Subject-Object or Human-Superhuman dualism, and thus being condemned to think in terms of power relations between the terms of these dualism. Though for purposes of *manipulating* them you may *have* to separate subject and object from each other and from language, remember that there are other purposes than manipulation." (Richard Rorty, *Essays on Heidegger and Others*, New York: Cambridge University Press, 1991, p. 35.)

philosophy.[30] This implies a reflected usage of language, i.e. an awareness of the fact that the language we use is a result of the past, thus our recent ways of speaking and thinking are not the only possible ones. Heidegger relying, then, considerably on the ancient Greek term *logos*, stresses that we are able to differentiate among things because language makes us capable of doing so. That is, language reveals the things all around and at the same time it shows up their presence or, as a preservative force, their lack of presence. In this way, we can comprehend the intimate relationship of *Being*, human beings, and things. Heidegger's view on language emphasizes that language is not a mere means for transmitting information, but rather a revelatory force, so to say a privilege of *Dasein*.

But everyday language is worn out, i.e. such language is unable to call forth *beings* and so conceals the original meaning of words. Thus, it is easy to forget the elementary force of language. And forgetting about this force provides an opportunity to think of language as a mere instrument for communication.[31]

Casting a glance at the task of poets and thinkers, I'll try to relate this to the issue of metaphors and images. "The simplicity of a rare richness speaks in the poem", wrote Heidegger.[32] "The language of the poem is a manifold enunciate-ing."[33] That is, we can say that the simplicity of a rare richness is a manifold enunciating. Adding to this that language reveals beings all around, the more-layer utterance of the poem is the richness of *Being*. If we interpret this from the point of view of the difference between the spoken and the not-spoken-aloud, we obtain Rorty's interpretation.

> The greater the ease with which we use that language, the less able we are to *hear* the words of that language, and so the less able we are to think of language as such. To think of language as such, in this sense, is to think of the fact that no language is fated or necessitated. So to forget the openness of beings is to forget about the possibility of alternative languages, and thus of alternative beings to those we know. It is in terms I was using before, to be so immersed in inquiry as to forget the possibility of poetry.[34]

---

[30] "Nevertheless, the ultimate business of philosophy is to preserve the *force of the most elemental words* in which Dasein expresses itself, and to keep the common understanding from leveling them off to that unintelligibility which functions in turn as a source of pseudo-problems." (Martin Heidegger, *Being and Time* (1927), transl. by John Macquarrie & Edward Robinson, Oxford: Basil Blackwell, 1962, p. 262.)

[31] Cf. Mallarmé's ideas about language becoming empty. See O. B. Hardison, *Disappearing*, p. 153.

[32] "Im Gedicht spricht das Einfache eines seltenen Reichtums." (M. Heidegger, *Nietzsche*, vol. II., Pfullingen: Günter Neske, 1961, p. 248.)

[33] Martin Heidegger, "Language" (1960), in: *Poetry. Language, Thought*, transl. by Albert Hofstadter, New York: Perennial Classics, 2001, p. 195.

[34] R. Rorty, *Essays on Heidegger and Others*, p. 45.

So we can see that this honourable role of poetry is based on the force of metaphor. And metaphor works differently than propositional statements. As Donald Davidson formulates it, "When we try to say what a metaphor 'means', we soon realize that there is no end to what we want to mention."[35] As well-known the kinship of metaphors and pictures, as questionable the possibility of the perfect transcription of metaphors and pictures to propositions.

If we want to interpret images in a hermeneutic way, metaphor seems to be the only mediator between the world of words and that of images. However, this claim can be opposed to the traditional perspective of the history of arts, since the latter bears the marks of the literate, as opposed to the oral, mind, i.e. it tries to verbalize what pictures show, tries to assimilate pictorial meaning to the conceptual framework of *logos* in a narrower sense. Gottfried Boehm criticized this attitude as one that leaves the difference between verbal and pictorial expression unreflected. Boehm emphasizes the one-sided optic of Western thinking based on the dominance of *logos*.

The Heideggerian critique of the historical change of *logos* suggests a kind of return to a less speculatively differentiated world – to the world of metaphors as it were. This attitude is, to some extent, in harmony with the demand to accept a world where there is more place for a receptive mental disposition.[36]

Through this brief survey of the history of communications technology in connection with a change in mental disposition, I have tried to shed light upon the tendency of the exhaustion of the conceptual framework which has evolved due to the distinctive force of *logos* in close relation with the change of communications technology and in accordance with the demands for clear-cut rationality. The recently emerging demand for a new kind of conceptual frame of reference seems to conquer the realm of images, which drives us towards *logos* as metaphor.

---

[35] *Ibid.*, p. 14 (the quote is from D. Davidson, *Inquiries into Truth and Interpretation*, 1984).

[36] Here I have in mind the Heideggerian notion of *Gelassenheit* – openness toward things.

Language, Myth, and Communication*

Difficulties closely related to language, such as the exactness of propositions, correctness of syllogisms, and the problem of avoiding misunderstandings, emerged quite early in literate history. This orientation necessarily led to the attitude that language can be grasped by way of grammatical, logical rules, by differentiation between sign and reference, etc. But there was also a non-formalizable, indeed hardly comprehensible, idea of language which was based upon the pre-Socratic concept of *logos*. These two different traditions were so dissimilar that there appeared to be no way to treat them within a common framework. However, I will attempt to show that these two traditions – i.e. the logical-grammatical approach, and the idea of language based on the pre-Socratic concept of *logos* – can be merged if language is regarded from the viewpoint of communication. The close relationships between the ancient concepts of *mythos*, *logos*, and a ritual point of view of communication provide the ground for this kind of approach to language. The communicational approach seems promising because it enables us to clear up the obscurities around the concept of language understood as *logos* (I have in mind Heidegger's concept of language in particular), and to recognize that language as an instrument for expressing ideas can indeed have a function in terms of cultural presuppositions and commitments.

I would first like to cast a glance at the earliest questions related to language, and then at the pre-Socratic concept of *logos* and myth (here I would expound the Heideggerian comprehension of *logos*). Then, I will outline the notion of communication, i.e. recall James Carey's distinction between the standard concept and the ritual view of communication. Finally, by combining the idea of communication in the ritual sense and the pre-Socratic idea of *logos*, I shall attempt to show that language, even when considered a mere instrument for transmitting information, is also capable of mediating commitments, faiths, and beliefs.

In spite of the fact that we can use language almost without reflection, it nonetheless became a target of investigation for Plato. In the dialog of *Cratylus*, we can read about whether the connection between names and things is conventional or is in accordance with nature. In the *Seventh Letter*, Plato describes the name of a thing as the first step towards knowledge of it.[1] In the

---

* First published in Jeff Bernard and Gloria Withalm (eds.), *Angewandte Semiotik 18/19. Mythen, Riten, Simulakra. Semiotische Perspektiven*, Wien: ÖGS 2001, vol. I, pp. 491-502.

[1] "For everything that exits there are three classes of objects through which knowledge about it must come; the knowledge itself is a fourth, and we must put as a fifth entity the actual

Sophist, Plato outlines the difference between verb and noun, and speaks about the first and the smallest talk (likely referring to sentences) and its falsehood/veracity.[2] Though Aristotle never used the term, 'logic',[3] he can be considered as the creator of logic as we understand it today. Aristotle's investigations of cognition with regard to the capability of comprehending the nature of things are collected under the title *Organon*.[4] Aristotle did not consider his logical instructions as science. He regarded them, as the title *Organon* shows, to be an instrument that everyone who wants to deal with science needs. This instrument helps to decide which propositions need verification and what kind of verification is suitable.

In the Middle Ages, several works were written about grammar,[5] rhetoric, and dialectics. These writings partly elaborated the ancient inheritance of Aristotelian thought, and partly defined the ontological status of general concepts as the decisive *universalia*-debate of that time required. During the modern age, language still remained an object of reflection. Descartes regarded language as a distinctive characteristic of men as compared to automatons. Several scholars called attention to false or erroneous uses of language.[6] But all

---

object of knowledge which is the true reality. We have then, first, a name, second, a description, third, an image, and fourth, a knowledge of the object... Of all these four, understanding approaches nearest in affinity and likeness to the fifth entity, while the others are more remote from it." (Plato, *Seventh Letter*, 342b-d, transl. by L. A. Post.)

[2] Plato, *Sophist*, 261 e – 264 b. Another meaning of *logos* appears here, viz. sentence. See W. Kneale – M. Kneale, *The Development of Logic*, Oxford Univ. Press, 1962, chapter I/5.

[3] Alexandros of Aphrodisias (3rd century) was the first who used *logiké* in the sense of modern logic. Sir David Ross calls our attention to the fact that even the term *logica*, which was never used before the time of Cicero, means dialectics. Regarding the difference between logic and dialectics, see Sir D. Ross, *Aristotle*, London: Routledge, 1995, p. 21, and also Kneale and Kneale, *The Development of Logic*.

[4] *Prior* and *Posterior Analytics*, *Topics* and *Sophistic Elenchi*, *Categories* and *De Interpretatione*. See Ross, *Aristotle*, chapter 2, and Kneale and Kneale, *The Development of Logic*, chapter 2.

[5] It is important to note that there is a difference between logic and grammar. The main difference, as Aristotle defined it, is that logic investigates the cognition of which signs are words. See Ross, *Aristotle*, chapter 2. Isidor of Seville wrote on grammar, rhetoric, and dialectics. He supposed that the knowledge of things makes it necessary to know their linguistic counterparts; Peter of Spain differentiated *significatio* and *suppositio*. The Renaissance gave birth to grammatical studies by Lorenzo Valla, Juan Luis Vives, and Peter Ramus, to mention a few famous opuses of that time.

[6] To name a few: Francis Bacon considered the idol of the market – according to which erroneous conceptions appear in language – the most troublesome; in the 4th chapter of *Leviathan*, Thomas Hobbes calls attention to misuses of language; Berkeley, in the preface of *A Treatise Concerning the Principles of Human Knowledge*, warns about deception and delusion due to language; and Leibniz, in his *New Essays on Human Understanding*, wrote about the perplexities of language.

these discussions also entail that language was implicitly presupposed to be a mere instrument of transmitting information. John Locke wrote in *An Essay Concerning Human Understanding*:

> The comfort and advantage of society not being to be had without communication of thoughts, it was necessary that man should find out some external sensible signs, whereof those invisible ideas, which his thoughts are made up of, might be made known to others.[7]

Even this sketchy and not-at-all-complete list shows that the investigation of language as an instrument for articulating ideas was prevalent. It implies a desire to avoid obscurities and ambiguities due to the use of language, and a need to carry on with Aristotelian logic, i.e. to make the right way of cognition explicit.

Interest in language increased in the 20th century:

> The scope of disciplines interested in language ranges from sociology to linguistics proper, from theology to cybernetics, from anthropology to fine arts; moreover these disciplines study language from a multitude of perspectives, such as grammatical, syntactical, semantical, logical, morphological, etc.[8]

Another characteristic of this century is that the communication between different traditions (i.e. the logical-grammatical approach and the idea of language based on the pre-Socratic notion of *logos*), is very sparse and scarcely successful. What can be the reason for this?

Throughout the history of philosophy, the comprehension of language was determined more and more by an instrumental point of view, i.e. from quite early on, the exactness, veracity, and provability of a proposition became the main concern of investigations. However, the fragments from Heraclitus suggest a different emphasis. According to David Wiggins, since Heraclitus couldn't possess a theory of logic, he couldn't think it to be the foundation of knowledge

---

[7] John Locke, *An Essay Concerning Human Understanding*, Oxford: Clarendon Press, 1984, vol. II., p. 8. "For Locke ... language was a complex vehicle of truth, 'the great bond that hold society together, and the common conduit, whereby the improvements of knowledge are conveyed from one man and one generation to another'." (Robert Pattison, *On Literacy. The Politics of the Word from Homer to the Age of Rock*, New York: Oxford Univ. Press, 1982, p. 141.)

[8] G. F. Sefler, *Language and the World. A Methodological Synthesis Within the Writings of Martin Heidegger and Ludwig Wittgenstein*, Atlantic Highlands, N. J.: Humanities Press, 1974, p. xix. Compare also "the fabric of our age is to a large extent language-textured" (*ibid.*, p. 195). See also E. Gellner's idea expressed in "The Crisis in the Humanities and the Mainstream of Philosophy" (in: *Crisis in the Humanities*, J.H. Plumb (ed.), London: Cox & Wyman Ltd., 1964) where he reflected on how and why language came into the focus of 20th-century philosophy. The deeper cause of the phenomenon he believed to be a crisis of the ability to grasp reality verbally.

and philosophy. Heraclitus was, rather, "a thinker best seen as relying on the language itself (not on a philosophy of logic or language or some theory of names or reference or predication) to fix the meaning of what he says".[9]

According to one of the Heraclitus fragments, it is possible to know things and manifest phenomena of nature although "[t]he real constitution is accustomed to hide itself".[10] If we can be as opened as the Delphic Apollo, we can gain exemplary patterns about the work of the cosmos.[11] Here, language has a very important role, i.e. due to language, the interaction between mind and things becomes possible.

> Of the Logos which is as I describe it men always prove to be uncomprehending, both before they have heard it and when once they have heard it. For although all things happen according to this Logos men are like people of no experience, even when they experience such words and deeds as I explain, when I distinguish each thing according to its constitution and declare how it is; but the rest of men fail to notice what they do after they wake up just as they forget what they do when asleep.
> Therefore it is necessary to follow the common; but although the Logos is common the many live as though they had a private understanding.
> Listening not to me but to the Logos it is wise to agree that all things are one.[12]

These three fragments make clear that all things function according to *logos*, and *logos* is common to everyone. For this reason, the world as well as *logos* are not strange to human beings. According to the last fragment quoted above, *logos* becomes manifest suggests Heraclitus by him himself being instructed by it – what Heraclitus says is in fact a manifestation of the *logos*.[13]

Heraclitus characterizes people who can not comprehend the connection between different phenomena as deaf. Their inability to gain such comprehendsion originates from a lack of understanding of how their own language works,

---

[9] David Wiggins, "Heraclitus' concept of flux, fire and material persistence", in M. Schofield – N. C. Nussbaum (eds.), *Language and Logos*, Cambridge and New York: Cambridge University Press, 1982, p. 28. Wiggins emphasizes the affinity between the Milesians and Heraclitus. As regards *logos*, cf. M. Heim's daring interpretation of Anaximander's notion *logos*. *Logos* is able to order chaos, "Logos is the emergence of identity out of the chaos of an infinite matrix of possibilities." (M. Heim, *Electric Language: A Philosophical Study of Word Processing*, New Haven and London: Yale University Press, 1987, p. 30.)

[10] G. S. Kirk, J. E. Raven and M. Schofield, *The Presocratic Philosophers*, Cambridge: Cambridge University Press, p. 192 (Fragment 123).

[11] "The lord whose oracle is in Delphi neither speaks out nor conceals, but gives a sign." (Kirk, Raven, Schofield, *The Presocratic Philosophers*, p. 209 [fragment 93].)

[12] *Ibid.*, p. 187 (fragments 1, 2, and 50).

[13] The obscurity of Heraclitus can be alleviated: Heraclitus thinks that the meaning of the words is not accidental. For instance, *bios* means life and bow. In that light, the next fragment seems to be less obscure: "The name of the bow is life but it works as death." (Quoted by Wiggins, in Schofield and Nussbaum (eds.), *Language and Logos*, p. 29 [Fragment 48].)

as if they were foreigners orincapable of hearing. Heraclitus believes that if human beings use their language only by routine, then there is no way to eliminate this kind of incapability.

> Yet even this does not mean that the human conditions is simply hopeless, closed upon its own hopelessness. What determines the identity of concepts and attaches common nouns to their denominata is what men *always did* – even before some men awake from their deafness to their own language. Practice is the anchor...[14]

The many ways of translating *logos* show some ambiguity. Even considering just the fragments from Heraclitus, it is apparent that several words are needed: language, word, speech, reason, intellect, explanation, proportion, and necessary structure.[15] These translations show that *logos* means a kind of ruling principle on the one hand, and a sort of articulation of it on the other. Actually, the meaning of *logos* is ambiguous even in Plato. Here, we can think of the unanswered question in the *Cratylos*, the true knowledge written in the soul, the difference between true knowledge and certainty and the role words play in it, i.e. "names do reveal the nature of the thing named".[16]

The Heideggerian interpretation of *logos* harmonizes with the Heraclitus fragments with regard to the holistic meaning and indeterminateness of *logos*. Talk[17] is the primary meaning of *logos* in *Being and Time*, but this translation needs further elaboration.[18] Logos means the process of making-manifest or letting-be-seen. "Through the articulate-ness, sc. Logos, of There-being, '...Total Meaningfulness ... *comes into words*...' "[19] In other words:

---

[14] Wiggins in Schofield and Nussbaum (eds.), *Language and Logos*, p. 31.

[15] See M. Heim, *Electric Language*, p. 34.

[16] J. Annas, "Knowledge and Language: the *Theaetetus* and the *Cratylus*", in: Schofield and Nussbaum (eds.), *Language and Logos*, p. 106.

[17] See M. Gelven, *A Commentary on Heidegger's Being and Time*, New York: Harper & Row Publishers, 1970, p. 103. However, we should note that other translations, using "discourse" instead of "talk", exist too.

[18] Heidegger's term is *Rede*. "Discourse 'lets something be seen' απο...;: that is, it lets us see something from the very thing which the discourse is about. In discourse (αποφανσιζ), so far as it is genuine, *what* is said [was geredet ist] is drawn *from* what the talk is about, so that discursive communication, in what it says [in ihrem Gesagten], makes manifest what it is talking about, and thus makes this accessible to the other party. This is the structure of the λογοζ aş αποφανσιζ... When fully concrete, discoursing (letting something be seen) has the character of speaking [Sprechens] – vocal proclamation in words." (M. Heidegger, *Being and Time* (1927), transl. by John Macquarrie & Edward Robinson, Oxford: Basil Blackwell, 1962, p. 56.) About the difficulties of translation see W. J. Richardson, *Heidegger. Through Phenomenology to Thought*, The Hague: Martinus Nijhoff, 1967, pp. 66 f.

[19] Richardson, *Heidegger*, p. 67.

> *logos* [i.e. disclosure], in contra-distinction to the usual meaning of the word, is the constitutive movement, and language is merely the way in which *logos* gets expressed. Language is that through which the *logos* makes itself mundane; through language it becomes an element of the world and can be treated like other things found in the world.[20]

"Letting-be-seen" or "making manifest" is the primary meaning of *logos* for Heidegger. All other translations – such as reason, judgement, concept, definition, ground and relationship[21] – are derived from this primary meaning. Heidegger emphasizes the danger of these derivations since they can lead us to lose the basic, primordial meaning, i.e. the fact that *logos* is a constitutive element of our world becomes concealed.

The definitions of logos, talk/discourse, and language are closely related to each other and their connection to the truth of Being is equally close. The manifestation of logos as felt comprehensibility (befindliche Verständlichkeit),[22] is talk. Language is the utterance of talk.

Heidegger does not deny that language can be an object of theoretical investigation as the various phenomena of nature are for science, i.e. language can be a present-at-hand (*vorhandenes*) thing. Language can be seen as ready-at-hand (*zuhandenes*), through which we articulate our comprehension of Being-with-others in everyday life. In this case, language is "a tool to be used in social intercourse",[23] a means of communication. But language is also a constitutive element of Being-in-the-world as it creates an important momentum of revealing. It is important to remind ourselves of the fact that in *Being and Time*, Heidegger tried to reconstruct the history of metaphysics, to reveal distortions, and to re-ask forgotten questions. Metaphysics gradually forgot to ask about the meaning of *Being*, i.e. the truth of *Being*.[24] The difference between *Being* and beings has vanished, the attitude towards the world became one-sidedly analyzing and theoretical, and the fact that humans are immersed in the world has also been forgotten.

---

[20] S. Corngold, "Sein und Zeit: Implications for Poetics", in W. V. Spanos (ed.), *Martin Heidegger and the Question of Literature*, Bloomington: Indiana University Press, 1979, p. 106.

[21] "Λογοζ gets 'translated' (and this means that it is always getting interpreted) as 'reason', 'judgement', 'concept', 'definition', 'ground', or 'relationship'. " (M. Heidegger, *Being and Time*, p. 55.)

[22] "The intelligibility of Being-in-the-world – an intelligibility which goes with a state of mind – *expresses itself as discourse*." (M. Heidegger, *Being and Time*, p. 204.)

[23] J. Aler, "Heidegger's Concept of Language in *Being and Time*", in: J. J. Kockelmans (ed.), *On Heidegger and Language*, Evanston: Northwestern University Press, 1972, p. 50.

[24] M. Heidegger, "Einleitung zu 'Was ist Metaphysik'", in: *Wegmarken* (*GA.9*), Frankfurt a. M.: Vittorio Klostermann, 1976, p. 377.

Truth gains its meaning in regard to revelation and Dasein.[25] "'Being-true' ('truth') means Being-uncovering."[26] And some paragraphs later: "Uncovering is a way of Being for Being-in-the-world... Disclosedness is constituted by state-of-mind, understanding, and discourse, and pertains equiprimordially to the world, to Being-in, and to the Self."[27]

Heidegger's later writings emphasize the constitutive character of language. We can read several passages about how we come to word concealment and un-concealment by *Sage* (myth, speaking).[28]

The most important characteristic of word and *Sage* is that it lets concealment and un-concealment, the disclosure and un-disclosure become visible. Heidegger defines the ancient Greek term *alétheia* as disclosure, particularly the disclosure of the relationship between humans and beings.[29] The essence of word and speaking is based on the essence of truth. "But man is not only a living creature who possesses language along with other capacities. Rather, language is the house of Being in which man ek-sists by dwelling, in that he belongs to the truth of Being, guarding it."[30]

Heidegger's ideas regarding the history of philosophy inform us that the evolution of philosophy has been, since nearly its beginning, a determined process even if the present state of affairs seems to be dangerous and filled with difficulties. Along the history of metaphysics the translation (i.e. interpretation) of *logos* and *aletheia* led quite far from the original comprehension of *Being* and language. This change can be seen in the alteration of terminology – *logos* becomes logic, and language becomes simply a tool for the exchange of ideas.

Heidegger's and Heraclitus's understanding of *logos* harmonize. There are some similarities in the case of the term *barbaroi*, as well. According to Heidegger, the Greek term *barbaroi* means myth, *logos*, and epos. For ancient Greeks, the difference between people becomes evident by word and saying. Those nations and people whose words are strange, not the known myth, *logos*,

---

[25] About the translation of *Dasein* see Gelven, M., *A Commentary*, pp. 22 f.

[26] M. Heidegger, *Being and Time*, p. 262.

[27] M. Heidegger, *Being and Time*, p. 263.

[28] "Überall west anfänglich das Aufgehen in das Unverborgene und das Untergehen in die Verbergung. Diese Wesung des Entbergens und Verbergens kommt jedem Seienden vorauf und überallhin und stets in allem Sein zum Wort und ins Wort der Griechen. Es ist das anfänglich Gesagte – die Sage." (M. Heidegger, *Parmenides*, Frankfurt am Main: Vittorio Klostermann, 1982, p. 99.)

[29] "[das] Wesen der αληθεια, die als Wesen des Seins selbst das Wesen des Menschen für sich in den Anspruch nimmt als dasjenige 'Wesen', das zum Seienden als einem solchen sich verhält." M. Heidegger, *Parmenides*, p. 102.

[30] M. Heidegger, "Letter on Humanism" (1947), in D. Farell Krell (ed.) *Basic Writings – Martin Heidegger* (revised and expanded edition), London and New York: Routledge, 2007, p. 237.

and epos, i.e. whose relation to *Being* is different from their own, were called *barbaroi*. The opposite concept to *barbaroi* is not "culture" or "education", but "staying in" the *logos* or myth.[31]

According to one of Heraclitus's fragments, "Evil witnesses are eyes and ears for men, if they have souls that do not understand their language." [32] As Kirk, Raven, and Schofield emphasize, "Here 'barbarian souls' are those that cannot understand the language of, cannot correctly interpret, the senses, but are misled by superficial appearances."[33]

Accordingly, the above-quoted fragment might be interpreted as referring to people who are not able to understand *logos* which is "common", and are unable to experience the cosmos as it has been arranged by *logos*, i.e. they are unable to interpret information gained by their senses in a correct way. *Logos* is determinative in experiencing our surroundings.

Finally, I would like to cast a brief glance at the concept of communication. I rely on James Carey's considerations, rooted in the pragmatic tradition. There is considerable convergence between the above-mentioned tradition and the ancient comprehension of "barbarian". Carey differentiates two notions of communication: on the one hand communication in the sense of passing, transmission and transfer, and on the other hand the ritual view of communication. The former mirrors the everyday use of the term. According to this idea, communication might be described as "'sending', 'transporting', or 'giving' information to others".[34] It is based on the metaphor of transportation in the sense of geography. It can be traced back to the consideration that in the interest of control, the transmissions of different goods and information are of the same process. And it holds true if we consider it historically until the appearance of telegraph: from the point of view of power and control, there is no essential difference between the motion of goods, people, or information. The invention and spread of the telegraph breaks down the identity of the concept of communication and transportation. In the 19th century, the superiority of communication over transportation had already become explicit. The telegraph is capable

---

[31] "Die Griechen unterscheiden sich gegen die anderen Menschentümer und Völker und nennen dieseą βαρβαροι, solche, die ein fremdartiges Wort haben, das nicht μυθοζ, das nicht λογοζ und nicht εποζist. Der Gegesatz zur 'Barbarei' ist für die Griechen nicht die 'Kultur', sondern das Innestehen im μυτοζ und λογοζ." (M. Heidegger, *Parmenides*, p. 103.) In another of Heidegger's writings, he takes *barbaroi* into account from a different point of view, viz. the ability to question is the difference between *barbaroi* and non-*barbaroi*. See Heidegger, M., *Die Frage nach dem Ding* (1962), Tübingen: Max Niemeyer Verlag, 1987, p. 32.

[32] Kirk, Raven, Schofield, *The Presocratic Philosophers*, p. 188 (Fragment 107).

[33] *Ibid.*

[34] J. Carey, *Communication as Culture. Essays on Media and Society*, New York and London: Routledge, 1992, p. 15.

of communicating ideas, which is the argumentation for the superiority of communication. Yet the metaphor lives on: communication is generally considered a process which disseminates information for the interest of power and control. The religious overtone connected to communication – recall Morse, and the idea that transportation/communication had an important role in the extension of God's kingdom – is eliminated by secularization, and the technological character of communication gains importance.

To use communication in a ritual sense is not usual, though this meaning goes much farther back in time then the above one. "In a ritual definition, communication is linked to terms such as 'sharing', participation', association', fellowship' and 'the possession of a common faith'."[35] According to the etymology of communication, the Latin roots are *communico, -are*, i.e. to do something with someone together, to make something public and to share something. So *communicatio* means a kind of imparting: I share with others in something that I would like to have in common with them. The fruit of this process is a kind of community, *communio*, establishing shared and common values.[36] The ritual view of communication does not mean an expansion of information in space, but it emphasizes the importance of maintaining the community in making common beliefs manifest. "It [the ritual view of communication] sees the original or highest manifestation of communication ... in the construction and maintenance of an ordered, meaningful cultural world that can serve as a control and container for human action."[37]

As Dewey put it:

> Men live in a community in virtue of things which they have in common; and communication is the way in which they come to possess things in common. What they must have in common in order to form a community or society are aims, beliefs, aspirations, knowledge – a common understanding – like-mindedness as sociologists say. Such things cannot be passed physically from one to another like bricks; they cannot be shared as persons would share a pie by dividing it into physical pieces. The communication which insures participation in a common understanding is one which secures similar emotional and intellectual disposition – like ways of responding to expectations and requirements.[38]

We can see that Dewey considers communication as having the ability to create and maintain community, similarly to Heraclitus's and Heidegger's understanding of *logos*. (I have in mind the special character of *barbaroi* in both

---

[35] *Ibid.*, p. 18.

[36] Kiss, U., "Információ és kommunikáció" (Information and Communication), in *Távlatok*, 1997/1, pp. 24-29.

[37] J. Carey, *Communication*, pp. 18 f.

[38] John Dewey, *Democracy and Education* (1916), Carbondale and Edwardsville: Southern Illinois University Press, 1985, p. 7.

cases.) The ancient concept of *logos* makes two very different comprehensions of language possible: one which tries to grasp it with regard to grammatical rules, regularities of cognition and operations, and their relations with elements of language, and the other which comprehends language from the point of view of its constitutive power. The latter comprehension seems to be fruitless – mainly from the other's point of view – as long as it doesn't have any matter of fact/graspable support. The former may be accused of an incapability of comprehending the whole of language or describing its real functioning since, for instance, language forms the object of the investigation independently of its use and users.

Casting a glance at language from the perspective of communication provides an opportunity to create a new frame of reference for investigation. Language as a very effective means of communication suggests a wider context of investigation, i.e. not involving exclusively signs and grammars, but rather the embedded usage of language. There is an inclination, at least to some extent, to label this perspective as a sociological point of view. However, it has (as Heidegger emphasized) metaphysical, and, we can add, epistemological relevance too. There are several new disciplines appearing (I have in mind psycho- and sociolinguistics), though not inside of the framework of philosophy. The ongoing change in the comprehension of language is rather in harmony with the ancient meaning of *logos*, though within a radically different conceptual framework.

PHILOSOPHY AND COMMUNICATIONS TECHNOLOGY

Changing Media: Perennial Challenge for Philosophy*

As the title of this paper suggests, it is about the history of philosophy as a history of searching for instruments and methods to avoid difficulties that are caused by the possibilities and limits of the ever new media of expressing ideas; difficulties dimly perceived, but not yet, at that moment, becoming fully conscious. First I am going to sketch some main characteristics of the major epochs of communications technology, then attempt to give a kind of outline of the history of philosophy with regard to its crucial problems. I'll focus especially on the changes of some concepts of metaphysics from the point of view of communications technology. By way of conclusion I would like to point out that it is not by chance that the technology of communication has today come into the focus of philosophical investigation.

Outline of a History of Communications Technologies: Epochs and Characteristics

The investigations especially of Harold Innis, Walter J. Ong, and Eric A. Havelock imply that the medium in which ideas are formulated is of major importance with regard to the cognitive capacity, intellectual disposition, and indeed the very organization, of communities. Let me here briefly compare the noetic worlds of oral and literate cultures.

A primary oral culture is one which – according to Ong's terminology – does not possess any knowledge of writing. Such a culture must be conservative and traditional. Since the only possibility to store information was in people's minds,[1] they developed a special language – a storage language as Havelock puts it – which was "devised orally for the purpose of survival".[2] This meant that it was possible to preserve information only by communicating, i.e. by sustaining live intercourse among people. This restriction required a special technology to weave ideas together, and to transmit awareness of the new facts of life. According to this special technology the way of expression was additive, redundant, and the expressions and words used were very closely embedded in

* First published in Kristóf Nyíri (ed.), *Mobile Learning. Essays on Philosophy, Psychology and Education*, Vienna: Passagen Verlag, 2003, pp. 69-79.

[1] Cf. Donald's emphasis on the novelty and importance of the written record, i.e. the external storage system: Merlin Donald, *A Mind So Rare: The Evolution of Human Consciousness*, New York: W. W. Norton & Company, 2001.

[2] E. A. Havelock, *The Muse Learns to Write. Reflections on Orality and Literacy from Antiquity to the Present*, New Haven and London: Yale University Press, 1986, p. 59.

concrete situations. Intercourse was empathetic, participatory and agonistic.[3] This kind of language was regulative, it strongly determined beliefs as well as people's attitudes and relationships to each other.

After the invention of alphabetical writing several modifications took place.[4] On the one hand the transmitting of ideas was liberated from certain restrictions; on the other hand written contexts had to be created for texts to be meaningful. With the help of writing, ideas became remarkably easier to handle and to elaborate. Part of the energy – earlier simply needed for remembering – could be mobilised for other purposes. Alphabetical writing gave rise to the technical-practical potential of creating concepts free of emotions, distantiated from the human life-world,[5] the potential to systematically analyze ideas, to regard events as linearly structured in time,[6] to be able to recognize the eternally human.[7] Knowledge could be transmitted in a new way, namely by written records.

While in a primarily oral community the preservation of factual knowledge and regulative norms was strictly bound to a perpetual process of face-to-face communication and a kind of embeddedness in concrete situations, due to alphabetical writing it became possible to preserve information without personal presence. Of course to replace the live situation with mute words required considerable intellectual effort and caused plenty of difficulties. The invention of the printing press made these changes more radical. Print, as Ong puts it, "goes farther in suggesting self-containment. Print encloses thought in thousands of copies of a work of exactly the same visual and physical consistency."[8]

The technology of printing texts in thousands of identical copies gave rise to the important novel concepts of originality and authorship. And let us note

---

[3] See Walter J. Ong, *Orality and Literacy. The Technologizing of the Word*, London and New York: Methuen, 1982, pp. 36-46. In the present volume see "The Tradition of *Logos* and a New Intellectual Attitude", pp. 16 f.

[4] Alphabetical writing was the instrument of the most accurate and the most abstract recording of the acoustic phenomenon of speech. Havelock underlines the fact, that – with respect to social control and governance – alphabetical writing was the only system to create a flexibility comparable to oral communication, since it did not ritualize and simplify the contents of the latter. See E. A. Havelock, *The Muse Learns to Write*, p. 59; Ong, *Orality and Literacy*, p. 28, and on the different types of writing also see Ong, *Orality and Literacy*, pp. 85 f.

[5] See Ong, *Orality and Literacy*, especially pp. 31-57 and 103-112; E. A. Havelock, *The Greek Concept of Justice: From its Shadow in Homer to Its Substance in Plato*, Cambridge, MA: Harvard University Press, 1978; and E. A. Havelock, *The Muse Learns to Write*.

[6] See Ong, *Orality and Literacy*, p. 143; P. Gendola, "Punktzeit: Zur Erfahrung in der Informationsgesellschaft", in R. Wendorf (ed.), *Im Netz der Zeit*, Stuttgart: S. Hirzel, 1989.

[7] See I. Hajnal, "Európai kultúrtörténet – írástörténet", in Hajnal, *Technika, művelődés: Tanulmányok*, Budapest: 1993, p. 18.

[8] Ong, *Orality and Literacy*, p. 132.

that this development was of course enhanced by the possibility to multiply images.

History of Philosophy as a Mirror of the Technology of Communication

When considering questions which for centuries were decisive topics of philosophy, one can soon realize that these questions were conceived originally as answers to some entirely practical experiences of everyday life. The historical modifications of some central concepts of metaphysics and indeed of the defined task of metaphysics can furnish good instances of the connections between the technology of communication and certain topics of philosophy.

As the *Oxford Companion to Philosophy*, which I will here use as a kind of reference point, puts it: "Metaphysics is the most abstract and in some views 'high-falutin' part of philosophy"[9]; but – let us add – a central part as well. Metaphysics created the basis of secularized western thought. Metaphysics, as the *Oxford Companion to Philosophy* formulates, aimed to define "the features of ultimate reality, what really exists and what it is that distinguishes that and makes it possible".[10] The process of secularisation took form first in the realm of metaphysics, and from its strivings originate ontology as well as epistemology.

Let us single out some main turning points in the history of metaphysics. Aristotle thought of what later came to be called metaphysics as "a science which investigates being as being and the attributes which belong to this in virtue of its own nature".[11] In the Middle Ages metaphysics was closely bound to theology, and the question of transcendency vs. immanency with regard to the fundament of being constituted its central issue. By the time of the Renaissance metaphysics became a kind of guarantee of scientific knowledge, as opposed to theology, and by the 17th century the creating of some methodological foundations for scientific knowledge was considered as the main task of metaphysics. Hegel looked upon metaphysics as something essential with regard to the "pure Spirit". To quote the *Oxford Companion to Philosophy* again, Hegel "thought reason could certainly do what Kant thought impossible, leading to the

---

9 Ted Honderich, *The Oxford Companion to Philosophy*, New York: Oxford University Press, 2005, p. 590.

10 *Ibid.*

11 "There is a science which investigates being as being and the attributes which belong to this in virtue of its own nature. Now this is not the same as any of the so-called special sciences; for none of these others treats universally of being as being. They cut off a part of being and investigate the attribute of this part; this is what the mathematical sciences for instance do. Now since we are seeking the first principles and the highest causes, clearly there must be some thing to which these belong in virtue of its own nature." (Aristotle, *Metaphysics*, Book IV, 1, transl. by W. D. Ross)

idea of an identification of self and object".[12] He tried to create a new basis for metaphysics by positing a historical objectifying activity of the Spirit which made it possible to comprehend as an identity concept and reality, the idea of a thing and the thing itself, the subject and the object. In the 20th century metaphysics took form in some holistic theories of reality while there of course emerged anti-metaphysical theories as well.

As the *Oxford Companion to Philosophy* writes: "In contemporary philosophy problems of metaphysics often take the form of a trilemma concerning some large and important feature of our life or discourse: illusion, well-founded appearance, and fundamental reality."[13] But looking at the 25 centuries of metaphysics we may realize that metaphysics saw as its main task simply: grasping reality in the right way. To be able to do that it was indispensable for it to give a definition of reality and later to define the suitable/adequate way of knowing the same. These questions of metaphysics originated in literacy, i.e. thought alienated from the living situation.[14] Contemplating pre-Socratic philosophy we can notice that, as the *Oxford Companion to Philosophy* aptly puts, it "was meta-physical in character, although it was initially presented in a dress which made it sound more like physics".[15] Think of the different notions of *arkhé*. *Arkhé* was the name of entities or things from which everything was made at the beginning.

After the invention of alphabetical writing, external memory, i.e. the preservation of information outside the human mind, developed new potentials. It was a radical change. As we saw, in a primary oral culture knowledge of different facts, norms and rules was available only through live intercourse. As knowledge became attainable through written records as well as via communication with others the question of reality and the accessibility of reality became central issues. Think of Plato's criticism in *Phaedrus,*[16] where he considered true knowledge as something "graven in the soul" in contrast to the written text which is useful only as "a reminiscence of what we know".[17]

---

[12] *The Oxford Companion to Philosophy*, p. 592.

[13] *Ibid.*, p. 593.

[14] Cf. Havelock's famous thesis to the effect that Platonic ideas – abstract concepts – became possible only after the invention of alphabetical writing.

[15] *The Oxford Companion to Philosophy*, p. 590.

[16] "[E]ven the best of writings are but a reminiscence of what we know, and ... only in principles of justice and goodness and nobility taught and communicated orally for the sake of instruction and graven in the soul, which is the true way of writing, is there clearness and perfection and seriousness, and ... such principles are a man's own and his legitimate offspring" *Phaedrus* 278a (transl. by Benjamin Jowett).

[17] In *The Seventh Letter* we can find some remarks which seem to query even this advantage of writing: "There neither is nor ever will be a treatise of mine on the subject. For it does not admit of exposition like other branches of knowledge; but after much converse about the

The main questions of medieval metaphysics were also related to the issue of reality. Thinking about universals, the existence of God and the question of transcendency and immanency with regard to the foundation of being gave rise to the central questions of metaphysics. Considerations connected to these questions had an abstract and complicated conceptual background. Concepts created a specific hierarchical world far from everyday experience.

Metaphysics in the Modern Age turned to the question of the reliability of knowledge – of knowledge that originates from the past, or from the senses. This specific distrust can be connected to the perceiving of a change in the technology of disseminating ideas, *viz.* the technology of the printing press. Due to the technology of printing, the unreliability of manuscripts with regard to verbalized ideas as well as diagrams, illustrations and pictorial statements became conspicuous. To find the right manner of cognition, which means at the same time to create a firm basis for the sciences, became the most important task for metaphysics. Think of Cartesian doubt. As Descartes put it:

> [And similarly] I thought that the sciences found in books – in those at least whose reasonings are only probable and which have no demonstrations, composed as they are of the gradually accumulated opinions of many different individuals – do not approach so near to the truth as the simple reasoning which a man of common sense can quite naturally carry out respecting the things which come immediately before him.[18]

The Crusoe tradition, or atomistic individualism, as Ernest Gellner referred to it in his posthumous book *Language and Solitude*, started with Descartes. This tradition considered human knowledge as developing step by step due to the intellectual effort of the individual – as indeed the practice of intellectual work was done more and more in isolation,[19] and directed at specialized topics. Knowledge was considered as something which the individual gains by his own cognitive efforts and/or private experiences, and the acquisition of knowledge as something which is only possible through abstract, distinct, and well-defined concepts, in the course of some systematically ordered process.

The intellectual prestige of science increased at that time. The importance of the technology of communication with regard to the development of early-modern science is excellently discussed by Elisabeth Eisenstein. She emphasizes that scientific development was due to the relatively easy acquisition of

---

matter itself and a life lived together, suddenly a light, as it were, is kindled in one soul by a flame that leaps to it from another, and thereafter sustains itself ." (341d, transl. by J. Harward).

[18] *Discourse*, transl. by Elisabeth Haldane.

[19] Which of course was not independent of the historical trend towards silent reading. See József Balogh, *Voces Paginarium: Adalékok a hangos olvasás és írás kérdéséhez*, Budapest: Franklin Társulat, 1921. For more details, see the chapter "Objectifying Process or Spiritual Involvement?" in the present volume, esp. p. 8.

contemporary texts as well as those of the past, and the possibility of correctly multiplying/duplicating diagrams, all of which were made possible by the printing press.

The intellectual atmosphere of typographic culture is usefully represented by the views of Leibniz on the question of reality. He used the well-articulated abstract terminology of Scholastic philosophy to which he added his firm trust in scientific truth. Leibniz thought reality knowable through the senses, but only with the help of science. "Only indivisible substances and their different states are absolutely real"[20] – he wrote. These indivisible substances, i.e. monads constitute the actual world. "The States of a monad are called 'perceptions'. ... In its perceptions every monad mirrors every other monad."[21] But there's no interaction among them – as Leibniz puts it, "they have no windows".[22] Every monad perceives everything in the universe including ourselves. Unfortunately most of our perceptions are confused. For example, as Benson Mates writes, "our perception of a rainbow is a confused visual perception of a huge number of droplets of water; subliminally we see each one, but at the conscious level we are aware only of the arc of colour."[23] Leibniz called this a "phenomenon", but a well-founded phenomenon as opposed to illusions, dreams or hallucinations. As Mates puts it: "The Information is only gleaned via scientific laws that allow us to conclude, from the features of the phenomena, the structure of the world they express. Our problem, in short, is to discover the laws that relate appearance to reality."[24]

In the 17th and 18th centuries language was considered as a distorting element with regard to knowledge. We can think of Francis Bacon who considered the Idol of the market – due to which erroneous conceptions appear in language – the most troublesome; Thomas Hobbes in the fourth chapter of *Leviathan* called attention to misuses of language; Berkeley in the preface of A *Treatise Concerning the Principles of Human Knowledge* warned about the deception and delusion caused by language; Leibniz also dealt with this problem in his *New Essays on Human Understanding.* He worked on the project of a universal language. As Eco writes: "He looked for a sort of mathematical lan-

---

[20] "What is real [res] is either immutable, i.e. God, or mutable, i.e. created being [ens creatum]. A created being is either a substance, which can perdure through change in itself, or an accident, which can not." (B. Mates, *The Philosophy of Leibniz: Metaphysics and Language,* New York: Oxford University Press, 1989, p. 47) Aggregates of substances Leibniz includes under the heading of real things.

[21] Mates, *The Philosophy of Leibniz*, p. 37.

[22] *Ibid.*, p. 39.

[23] *Ibid.*, p. 41.

[24] *Ibid.*, p. 200.

guage by virtue of which scholars, when debating a problem, could sit around a table, to implement some logical calculuses, and find a common truth."[25]

The other consequence of the printing press which is in close connection with the miraculous development of science was of course the re-detection of pictures. As Eisenstein noted, it was at that time that there spread the old "Chinese maxim that a single picture was more valuable than many words".[26] And as Robert Pattison writes: "Leibniz considered Chinese characters as the prototype for an ideal philosophical system of notation, free of ambiguities of our spoken language and directly correlated to the objects of thought without the distorting intermediacy of spoken language."[27]

The demand to make language unambiguous with the help of a certain scientific method on the one hand, and the recognition of the expressive power of pictures on the other, can be considered as attempts which quite explicitly confront the barriers of thinking determined by literacy.

In the 19th century Nietzsche emphasized the magnificence of ancient poetry as against the pale, abstract and inanimate writings of his time.[28] In the 20th century there were many critical remarks concerning the metaphysics of the Modern Age and the accompanying notions of language. Also in that century several terms emerged in philosophy which were bound up with the idea of everyday practice. The issues e. g. of reality and of the nature of language now emerged from the perspective of everyday life and practice.

---

[25] "The first European scholar to speak of a 'universal character' was Francis Bacon *(De dignitate and augmentis scientiarum)* and, in order to prove its possibility, he quoted the Chinese writing. Curiously enough, neither Bacon nor Wilkins understood the iconic origin of ideograms, and took them as purely conventional devices: in any case the ideograms looked as endowed with the double property of being universal and able to establish a direct contact between the character and the idea. The impact of the discovery of Chinese ideograms had an enormous influence on the development of the research for a Universal Philosophical Language in Europe, but it is not this the point that interests us today. – Leibniz was working on binary calculus, that is that mathematical calculus which proceeds by 0 and 1 and which is still used today for programming computers. Leibniz was convinced that such a calculus had metaphysical foundation because it reflects the dialectic between God and Nothingness." (Umberto Eco, "From Marco Polo to Leibniz: Stories of Intercultural Misunderstanding", lecture given in 1996, http://www.italianacademy.columbia.edu/pdfs/lectures/eco_marco.pdf.)

[26] Elisabeth L. Eisenstein, *The Printing Press as an Agent of Change: Communications and Cultural Transformation in Early-Modern Europe,* I-II, Cambridge: Cambridge University Press, 1979, p. 69.

[27] Robert Pattison, *On Literacy: The Politics of the Word from Homer to the Age of Rock,* Oxford: Oxford University Press, 1984, p. 34.

[28] See *Nietzsches Werke,* Leipzig: Alfred Kröner, vol. XVIII, 1912, p. 248. Quoted by Balogh, *Voces Paginarium*, p. 27.

We can sum up these changes by saying that philosophical interest gradually turned its back on literacy alienated from everyday practice and tried to find new ways to grasp reality through coming to terms with the actualities of everyday life. This turn is occurring in the age of secondary orality. Secondary orality set in with telephone, radio, television, and the various kinds of sound tapes and similar electronic technologies. "This new orality", as Ong puts it,

> has striking resemblances to the old in its participatory mystique, its fostering of communal sense, its concentration on the present moment, and even in its formulas. But it is essentially a more deliberate and self-conscious orality based permanently on the use of writing and print, which are essential for the manufacture and operation of the equipment and for its use as well.[29]

To be sure, I do not think that all the said changes are exclusively caused by modifications in the technology of communication. But I do think that they are decisive elements with regard to the general features of intellectual capacity and attitude.

Let me call attention to some remarks by William M. Ivins, Jr. In his *Prints and Visual Communication* Ivins sketched the development of the printed picture from the 16th century to the emergence of photography in the 19th century. Besides the technical inventions he also takes into account their cognitive consequences. "I propose", writes Ivins, "to look at the evidence from the point of view of that communication of visual information and ideas which, for the last four centuries, has been the primary function of the exactly repeatable pictorial statement."[30]

The importance of exact reproduction becomes obvious if we think of verbal communication. Ivins stresses that the institution of verbal communication is built upon the fact of there being exactly repeatable linguistic elements. We consider exact repeatability as something evident in the case of oral or written communication, mediating thoughts verbally, but this is not the case with pictures. As Ivins puts it: "pictures were of little use as definitions or descriptions because they could not be exactly repeated" for a long time.[31] There is no use of definitions and descriptions which can change each time when they are repeated. As Ivins continued:

> Plato's Ideas and Aristotle's forms, essences, and definitions, are specimens of this transference of reality from the object to the exactly repeat-able and therefore seemingly permanent verbal formula. An essence, in fact, is not part of the object but part of the definition. Also, I believe, the well-known notions of substance and attributable

---

[29] Ong, *Orality and Literacy*, p. 136.
[30] W. M. Ivins, Jr., *Prints and Visual Communication*, Cambridge, MA: Harvard University Press, 1953, p. 24.
[31] *Ibid.*, p. 62.

qualities can be derived from this operational dependence upon exactly repeatable verbal descriptions and definitions – for the very linear order in which words have to be used results in a syntactical time order analysis of qualities that actually are simultaneous and so intermingled and interrelated that no quality can be removed from one of the bundles of qualities we call objects without changing both it and all the other qualities.[32]

Accordingly, ancient Greeks, and even "reasonable" people of the 19th century, lack the proper technology with which they can take into account, and handle, particularities; so they think in generalities."[33]

According to Ivins' remarks verbalization was tending to be abstract and general from the moment there was a proper technology to be able to repeat definitions correctly. Since the correct duplicating of pictures or pictorial statements became possible only in the 19th century, the main realm for thinking, the realm expected to be suitable for the accumulation of knowledge, was, in the beginning, oral discourse, and, after the invention of alphabetical writing, the written text.

But verbalization at the same time means some loss of information. Let me here recall some remarks by Michael Polanyi to the effect that in the course of reproducing or describing our experiences we have to choose between either being precise or being close to the richness of the living context. It is impossible to simultaneously grasp precision and the "immense wealth of living shapes", the richness of the whole context. As he put it:

> Higher degrees of formalization make statements of science more precise, its inferences more impersonal and correspondingly more "reversible"; but every step towards this ideal is achieved by a progressive sacrifice of content. The immense wealth of living shapes governed by the descriptive sciences is narrowed down to bare pointer-readings for the purpose of the exact sciences, and experience vanishes altogether from our direct sight as we pass on to pure mathematics. – There is a corresponding variation in the tacit coefficient of speech. In order to describe experience more fully language must be less precise. But greater imprecision brings more effectively into play the powers of inarticulate judgement required to resolve the ensuing indeterminacy of speech. So it is our personal participation that governs the richness of concrete experience to which our speech can refer. Only by the aid of this tacit coefficient could we ever say anything at all about experience – a conclusion I have reached already by showing that the process of denotation is itself unformalizable.[34]

---

[32] *Ibid.*, p. 63.

[33] See Ivins, *ibid.*, p. 91.

[34] M. Polanyi, *Personal Knowledge: Towards a Post-Critical Philosophy*, London: Routledge & Kegan Paul, 1962, pp. 86 f.

Similarly, the process of "applying language to things" is "necessarily unformalized".[35] – In the light of these remarks by Polanyi the situation described by Ivins becomes a little bit more complicated: while expressing ideas verbally is effective with regard to the accumulation and dissemination of knowledge, at the same time verbalization creates precisely those very unsolvable anomalies which philosophy traditionally tried to solve or to eliminate.

Why Is Communication a Significant Issue of Philosophy?

Gellner, in a critical remark on Wittgenstein, considers philosophy as an institution to devise proper concepts for the purpose of grasping reality.[36] His attitude furnishes a good instance of the traditional attitude of philosophy flourishing in the culture of literacy. And indeed philosophy practically from its beginnings is tied to literacy. In the course of the history of philosophy it is the first transition from primary orality to literacy which we can, many centuries later, realise to have happened at all, and this realisation quite probably hangs together with the next radical change now imminent.

Possibilities recently provided by the technology of communication make us capable to re-present and/or to communicate some events, some phenomena or some new experience almost as if they were observed by ourselves in a live situation. This means that it is not necessary to verbalize the whole situation of an experience or some insight – the pressure to prepare linearly ordered text is decreasing. As the pressure of having to interpret ideas in the very process of formulating them decreases, we can concentrate on the communication of them, i.e. the lesser the obstacle is to formulate ideas the more visible is the importance of the technology we use for communicating them.

---

[35] "The lesson derived [earlier in this book] from reflecting on the application of the exact sciences to experience can be extended now as follows. We have seen that in all applications of a formalism to experience there is an indeterminacy involved, which must be resolved by the observer on the ground of unspecifiable criteria. Now we may say further that the process of applying language to things is also necessarily unformalized: that it is inarticulate". (*Ibid.*, p. 81.)

[36] "But philosophy started precisely because commonsensical notions became unworkable or inadequate." (E. Gellner, "The Crisis in the Humanities and the Mainstream of Philosophy", in J. H. Plumb (ed.), *Crisis in the Humanities*, Middlesex: Penguin Books, 1964, p. 68.)

Virtual Realities – Old and New*

Virtual reality is at the focus of several research projects. Differing opinions have been articulated about this phenomenon: some think of it as something dangerous, while others emphasize the new horizons it can open. Rather than attempt to decide who is right, I will try to shed light on the connection between the concept of reality and the medium we use to articulate our thoughts.

As Myron W. Krueger states:

> [Virtual reality] is a technology that can be applied to every human activity and can be used to mediate in every human transaction. Since you are completely immersed in the virtual world, virtual reality constitutes a new form of human experience – one that may be as important to the future as film, theatre, and literature have been of the past. Its potential impact is so broad that it may define the culture that results from its use.[1]

Krueger calls attention to the technological movement/capacity of virtual reality, as well as its social and cultural consequences. I would like to say something about the cognitive and intellectual aspects of the technology of communication on the one hand, and to highlight the impact of communications technology on the concept of reality on the other hand. Accordingly, I will first roughly sketch the main characteristics of the major epochs of the technology of communication, then outline the changes of the concept of reality, and by way of conclusion, I will focus on reality as a social construction.

Preliminary Remarks

In one of his papers, James Carey calls attention to the close relation of communication, cognition, and community.

> Our minds and lives are shaped by our total experience – or better, by representations of experience and, as Williams has argued, a name for this experience is communication. If one tries to examine society as a form of communication, one sees it as a process whereby reality is created, shared, modified, and preserved.[2]

Carey's statements seem to be correct, and I will subsequently argue in favour of his ideas. By examining the results of investigations by Harold Innis,

---

* First published in Jeff Bernard and Gloria Withalm (eds.), *Semiotische Berichte. Special Issue: Media – Old and New*, 2003/1-4, pp. 197-210.

[1] Michael Heim, *The Metaphysics of Virtual Reality*, New York and Oxford: Oxford University Press, 1983, foreword by Myron W. Kruger, p. vii.

[2] James W. Carey, "A Cultural Approach to Communication", in: *Communication as Culture. Essays on Media and Society* (1989), New York, London: Routledge, 1992, p. 33.

Walter J. Ong and Eric A. Havelock, it becomes apparent that the medium of formulating ideas has considerable impact on both thinking and the organization of communities. Just consider the main differences between oral and literate societies with regard to cognitive capacity, intellectual disposition, and the organization of community. As discussed in previous chapters, in primary oral cultures there are considerable restrictions because of the capacity of memory, and, obviously, many different techniques to help memorization. Due to the limits of memory, there is not much room for change.[3] The possibility of recording speech opened the horizon for abstract and complicated thinking. Demand for abstract and general nouns emerged, and the emphasis from the actual situation moved towards general statements and truths.[4]

Within the framework of literacy, it is crucial to mention the emergence of the printing press. The opportunity to create identical duplicate copies of written records made the changes caused by alphabetical writing more radical. As Ong summarized these changes:

> Before print, writing itself encouraged some sense of noetic closure. By isolating thought on a written surface, detached from any interlocutor, making utterance in this sense autonomous and indifferent to attack, writing presents utterance and thought as uninvolved with all else, somehow self-contained, complete. Print in the same way situates utterance and thought on a surface disengaged from everything else, but it also goes farther in suggesting self-containment. Print encloses thought in thousands of copies of a work of exactly the same visual and physical consistency.[5]

Reality Beyond Experience

Considering the main characteristics of philosophical thinking, we may notice an increasing trust in reason, a certain dominance of the perspective of the human subject as observer, and a strengthening insistence on method, all after the age when cursive handwriting became widespread and then when printing was accepted.

To make the connection between intellectual attitude/orientation and the media by which thoughts are expressed more evident, allow me to recall some ideas from Michael Polanyi and William M. Ivins, Jr.

---

[3] For more details on orality, see Walter J. Ong, *Orality and Literacy. The Technologizing of the Word*, London and New York: Methuen, 1982, pp. 36-46. In the present volume see the chapter "The Tradition of *Logos* and a New Intellectual Attitude", pp. 16 f.

[4] For more details on literacy, see E. A. Havelock, *The Muse Learns to Write. Reflections on Orality and Literacy from Antiquity to the Present*, New Haven and London: Yale University Press 1986, p. 59; Ong, *Orality and Literacy*, p. 28; and about the different forms of writing *ibid.*, p. 85. In the present volume see the chapter "Tradition of *Logos* and a New Intellectual Attitude", pp. 17 f.

[5] Ong, *Orality and Literacy*, p. 132.

First, I'll refer to some remarks by Michael Polanyi. Polanyi calls attention to the point that it is not possible to be precise and rich in detail at the same time. It is hardly possible to perfectly describe the "immense wealth of living shapes". Because "our personal participation that governs the richness of concrete experience to which our speech can refer" is necessary for such a description, precision is damaged. However, the degree of precision is not absolute, since "the process of denotation is itself unformalizable". That is, even if we define something with caution, the application of this definition at each time is not prescribed.[6]

As we can see, Polanyi illuminates the limits of verbalization. Ivins pointed to the reason for the compulsion of verbalization. In *Prints and Visual Communication*, Ivins sketches the development of the printed picture from the 16th century to the emergence of photography in the 19th century. In his discussion of technical inventions, he takes into account their consequences in scientific theory. "I propose", writes Ivins, "to look at the evidence from the point of view of that communication of visual information and ideas which, for the last four centuries, has been the primary function of the exactly repeatable pictorial statement."[7] That is, exact duplicability is necessary because words work as definitions. Words as exactly repeatable items of speech or text are obvious, but in the case of images the situation is rather different. For a long time, "pictures were of little use as definitions or descriptions because they could not be exactly repeated".[8] The impossibility of exact duplication makes images useless as definitions and descriptions – they can change each time they are reproduced. Ivins clearly relates some of the main concerns of Western philosophy ("Plato's Ideas and Aristotle's forms, essences, and definitions") to "the exactly repeatable and therefore seemingly permanent verbal formula."[9] Accordingly, in the absence of a proper technology with which particularities can be taken into account and handled, generalities determine the horizon of philosophical considerations. As Ivins wrote:

> They [thinking people of the 18th and 19th century] had not the means to think in particularities, which are always irrational, and they had to think in generalities… To a very considerable extent they were still in the situation and the frame of mind that had caused the Greeks to think as they did about some of the basic problems in philosophy. Thus just as the ancient Greeks developed the Platonic doctrine of Ideas and Aristotle's

[6] Michael Polanyi, *Personal Knowledge – Towards a Post-Critical Philosophy*, London: Routledge & Kegan Paul, 1962, pp. 86 f. For the quotation, see p. 45.

[7] William M. Ivins, Jr., *Prints and Visual Communication*, Cambridge, MA: Harvard University Press, 1953, p. 24.

[8] *Ibid.*, p. 62.

[9] *Ibid.*, p. 63 (for the quote, see the chapter above, "Changing Media: Perennial Challenge for Philosophy", pp. 44 f.).

essences, so the eighteenth century developed ideas of the Truth of Science and of the Laws of Nature.[10]

With Ivins pointing out that even the field of problems is determined by the technology of formulating ideas and Polanyi pointing to the difficulties of verbal representation, it becomes clear that the noetical world of literacy must be different from that of primary orality.

I'm now going to sketch the alteration of the concept of reality with regard to communications technology. First, I would like to outline the concept of reality by Duns Scotus which mirrors a well-internalized literacy, and then I will discuss the Lebnizian theory of reality as a good example of print culture.

Duns Scotus's notion of reality is a well-differentiated and quite complicated masterpiece of the 13th century. The phrase "virtuality" goes back to his concept of the primary object of the intellect.

> [Duns Scotus] argues that we have a concept of being that applies not analogically but univocally to all realities, including God, even though all our concepts originate from the sense experience of material beings. He argues also that the primary object of the human intellect is being in this universal sense – though he acknowledges that for some reason we are unable in this life to know directly any but material beings.[11]

To clarify this universal sense, recall Porphyry's tree.[12] Being is at the top of this tree. Below it there are 10 branches, *viz.* the ten Aristotelian categories. One of them, substance, is divided into material and immaterial substance. Below material substance are two subgenera: living and inanimate. Below living entities is a distinction between man and other species of animals. At last, below man there are branches towards individual persons. The "individuals of an ultimate species are differentiated from one another by what he calls individuating difference, or *Thisness*, in Latin *haecceitas*."[13] So we can define an individual entity with the genus it belongs to and *haecceitas*. But the difference that is meant by *haecceitas* – according to Duns Scotus – is not a real difference. Duns Scotus differentiates three kinds of difference: real, formal and rational.

> There is a distinction of reason when there is no difference except what the human intellect creates. There is a real distinction when, apart from any intellectual activity, there are two things (or at least two potential things – potential parts that can in some circumstances exist as thing and thing). There is a formal distinction when there is no

[10] *Ibid.*, p. 91.

[11] Kilcullen, R. J., *Later Medieval Philosophy – The Primary Object of the Intellect*, 1996 (Tape 5), http://www.humanities.mq.edu.au/Ockham/z3605.html.

[12] Porphyry (232-300) was a neo-Platonic philosopher. His "tree" had a great influence on medieval philosophy.

[13] Kilcullen, R. J., *Later Medieval Philosophy – The First Object of the Intellect*, 1996 (Week 4), http://www.humanities.mq.edu.au/Ockham/z3604.html.

real distinction, but there is, even before any activity of the intellect, two realities or formalities that can be grasped (when the human mind does arrive on the scene) in two concepts neither of which is part of the definition of the other. [14]

So this last distinction is not real – it is not a difference between two things – but it is already there somehow in the divine essence, waiting for the human mind to discover it.

We can approach *thing* (res) from a different direction, *viz.* from the concept of transcendentals. There are branches not only downwards but to the side as well. Here we can find transcendentals (because they are not contained within the supreme genera, i.e. being) such as *one*, *good*, *true*, and (the most important from the point of view of reality) *res,* which means thing. These transcendentals are coextensive, i.e. every being is thing, one, true, and good.

Being, transcendentals, individuating difference and accidents create quite a special interdependence. On the one hand being is common to all things, even to material substances. On the other hand being is not predictive of transcendentals and ultimate differences. But transcendentals are proper accidents of being, i.e. they are not the essence of it, not part of the definition, but an implication or effect of the characteristics that belong to the definition. Transcendentals and ultimate differences are contained *virtually* in being. What does is meant by "virtually"? If we think of the Porphyrian tree, "virtually contained" means to be found under what contains them, as a sub-heading, but not found in the concept of the subject when you spell out its definition.

According to Duns Scotus, the difference between real and virtual is the difference of actual, thing-like entities on the one hand and a kind of insight into the hierarchy described by the Porphyrian tree and a kind of perception of certain potentiality on the other hand. Everything that is actually accessible by the senses is real, as are things that have substantial being and/or whose existence is predicatable. What is accessible by knowledge and/or some kind of realization and that which is possible but not actual, Duns Scotus defined as *virtual*.

Clearly, the concept of reality by Duns Scotus implied literacy: without the technical invention of alphabetical writing, this sophisticated conceptual framework couldn't exist. And, of course, Duns Scotus's distinctions of real and virtual would be impossible without written records, since they require insights that are based on elaborated conceptual analysis. That is, all this is bound to literacy not only with regard to a sophisticated conceptual framework, it also needs to be capable of objectification for the further elaboration of notions.

[14] See R. J. Kilcullen, *The Primary Object of the Intellect*, http://www.humanities.mq.edu.au/Ockham/z3605.html.

On the question of reality, Leibniz nicely represents the intellectual atmosphere of typographic culture. Leibniz's trust in science mirrors the unique development of science in his time.

Elisabeth Eisentein sums up the new horizon opened up by the printing press as follows:

> The idea of observing natural phenomena directly and carefully was as old as Aristotle or Galen. The phrase 'ego vidi' was not lacking in many scholastic commentaries. Manuscript margins show skilful renderings of species of recognizable insects and birds. But the chance to discard inherited schemes, collect fresh data and build improved models on them came only after print. In view of the output of corrupted data ('of human books copied badly') during the first century of printing and in view of the new possibility of duplicating fresh records, a reaction of some new kind against accepted texts, fixed lectures and received opinion was almost inevitable.[15]

Leibniz believed reality is knowable through the senses, but only with the help of science. As he wrote: "Only indivisible substances and their different states are absolutely real." – And at another place one can read: "What is real [*res*] is either immutable, i.e. God, or mutable, i.e. created being [*ens creatum*]. A created being is either a substance, which can perdure through change in itself, or an accident, which can not.[16]

Leibniz includes aggregates of substances as real things. These indivisible substances, i.e. monads, constitute the actual world. "The states of a monad are called 'perceptions' ... In its perceptions every monad 'mirrors' every other monad."[17] But there's no interaction among them, "they have no windows".[18] Every monad perceives everything in the universe including ourselves. Unfortunately, most of our perceptions are confused. For example, "our perception of a rainbow is a confused visual perception of a huge number of droplets of water; subliminally we see each one, but at the conscious level we are aware only of the arc of colour."[19]

Leibniz refers to this as phenomenon, but a well-founded phenomenon opposed to illusions, dreams, or hallucinations. The possibility of confusion delimitates the information value about the external world.

---

[15] Elisabeth L. Eisenstein, *The Printing Press as an Agent of Change: Communications and Cultural Transformation in Early-Modern Europe*, vol. II. Cambridge: Cambridge University Press, 1979, pp. 475 f.

[16] Benson Mates, *The Philosophy of Leibniz. Metaphysics and Language* (1986), New York, Oxford: Oxford University Press, 1989, p. 47.

[17] *Ibid.*, p. 37.

[18] *Ibid.*, p. 39.

[19] *Ibid.*, p. 41.

In short, for Leibniz the rational man's perceptions do give him information about a world external to himself, but this is not because the world is qualitatively similar to his perceptions; rather, the information is only gleaned via scientific laws that allow us to conclude, from the features of the phenomena, the structure of the world they express. Our problem, in short, is to discover the laws that relate appearance to reality.[20]

According to Leibniz, a true picture of reality was only possible with the help of science. He used the distinct, abstract terminology of Scholastic philosophy supplemented with a firm trust in scientific truth and the hope of *Lingua Philosophica*, i.e. an artificial algorithmic language.

I won't touch upon Kant's transcendental philosophy as a well-known example of a special comprehension of reality. Instead, I'll try to sketch the consequences of recent new media on the concept of reality.

Reality as a Construct?

I would like to first outline the shift from literacy to secondary orality, and illustrate this change from the perspective of philosophy. Then, I will examine Peter Berger's and Thomas Luckmann's theory of reality as an example of the change of intellectual attitude.

As Ong summarized the main characteristics of secondary orality:

[W]ith telephone, radio, television and various kinds of sound tape, electronic technology has brought us into the age of 'secondary orality'. This new orality has striking resemblances to the old in its participatory mystique, its fostering of communal sense, its concentration on the present moment, and even in its formulas. But it is essentially a more deliberate and self-conscious orality based permanently on the use of writing and print, which are essential for the manufacture and operation of the equipment and for its use as well.[21]

And – we have to add – the significance of computer mediated communication has recently been increasing. The technological inventions listed by Ong mainly fostered the acoustical and personal element of verbalization; computer mediated communication has made other ways of formulating ideas possible, i.e. it became possible to communicate ideas with the help of multimodal means. The notions I'll be examining are in close connection with this kind of orality, inasmuch as the technological opportunities of secondary orality can intermediate the "immense wealth" of live situations as opposed to the strict, frozen, and abstract precision of literacy. The concept of reality in the age of literacy is an abstract creation, considered as something out of the reach of unqualified human intellect, i.e. reality is accessible only through hard

[20] *Ibid.*, p. 200.

[21] Ong, *Orality and Literacy*, p. 136.

intellectual effort. Or as in the case of Kant, reality creates the ontological foundation for the accessible phenomenal world. In the 20th century, instead of a distinct and complex notion of reality distanced from everyday practice, a reality emerged which is defined as something resulting from everyday life.

Heidegger takes a decisive step when he leaves the immanence of mind behind, and tries to focus on everyday life. The way he defines *Dasein* and its surroundings clearly shows that he considers human beings primordially and fundamentally immersed in the world.

> 'Reality', as an ontological term, is one which we have related to entities within-the-world. If it serves to designate this kind of Being in general, then readiness-to-hand and presence-at-hand function as modes of Reality... In order of the ways in which things are connected in their ontological foundations and in the order of any possible categorical and existential demonstration, *Reality is referred back to the phenomenon of care*. But the fact that Reality is ontologically grounded in the Being of Dasein, does not signify that only when Dasein exists and as long as Dasein exists can the Real be as that which in itself it is.[22]

The distinction of readiness-to-hand and presence-at-hand are different modes of being engaged in the world. Readiness-to-hand is primordial, i.e. it is fundamental to *Dasein* to cope with its environment. Relying on "referential totality", things emerged as ready-to-hand means.[23] Entities around us emerge as present-at-hand things in case of theoretical investigations or in case of "unusability".[24] Heidegger rejects Cartesian dualism, the separation of subject as *res cogitans* and the surrounding world as *res extensa*, he rather emphasizes the embeddedness into the world, which means at the same time different engagements.

Similarly, Heidegger considers language ready-to-hand, i.e. part of the world of *Dasein* which is a "*with-world*". "Being-in is *Being-with* Others."[25] Of course, language is decisive as regards revealing this world in which humans are immersed.

There is a deep similarity between Heidegger and Wittgenstein with regard to their intention to keep everyday practice in mind instead of theoretical considerations.

Wittgenstein calls attention to difficulties which arise when we think of language along the lines that Socrates understood it in the *Theaetetus*, i.e. when

---

[22] Martin Heidegger, *Being and Time* (1927), transl. by John Macquarrie & Edward Robinson, Oxford: Basil Blackwell, 1962, pp. 254 f.

[23] *Ibid.*, pp. 69 f.

[24] *Ibid.*, pp. 102 f.

[25] *Ibid.*, p. 155.

we consider "speech" as if it was "the composition of names"[26], distinguishing between composite and simple elements, essential and inessential features of an object,[27] and the like, which creates unsolvable anomalies. These anomalies can be eliminated only by a radical change of perspective. According to the *Philosophical Investigations*, language is best seen from perspective of its usage.[28] The "meaning of a word", writes Wittgenstein, "is its use in the language."[29]

Soon after this, Wittgenstein quotes from the *Tractatus*:

> (Tractatus Logico-Philosophicus, 4.5): "The general form of proposition is: this is how things are." – That is the kind of proposition that one repeats to oneself countless times. One thinks that one is tracing the outline of the thing's nature over and over again, and one is merely tracing round the frame through which we look at it.[30]

According to this paragraph, reality takes form along the usage of a language, i.e. to grasp reality in verbal expressions means, at the same time, to structure it. "'So you are saying that human agreement decides what is true and what is false?' – It is what human beings *say* that is true and false; and they agree in the language they use. That is not agreement in opinions but in form of life."[31] This paragraph shows language as a decisive element with regard to reality. According to Wittgenstein's notions on language, reality is outlined in the usage of language, i.e. in the form of life.

The next theory of reality I will examine embeds the above-sketched considerations into the framework of social/communal life. *The Social Construction of Reality*, the book by Peter Berger and Thomas Luckmann, was first published in 1966. The authors aimed to complete the theory of sociology of knowledge. As they put it: "a sociology of knowledge will have to deal not only with the empirical variety of "knowledge" in human societies, but also with the process by which *any* body of "knowledge" comes to be socially established as "reality"."[32]

---

[26] Ludwig Wittgenstein, *Philosophical Investigations* (transl. by G. E. M. Anscombe), Oxford: Basil Blackwell, 1963, § 46. For more details, see the chapter below, "Solitude, Culture, and the Technology of Communication", especially on p. 85.

[27] *Ibid.*, § 62.

[28] The same basic conviction of Wittgenstein's is of course present in those remarks which deal with the function or limits of philosophy. See *ibid.*, §123, §124, and §125.

[29] *Ibid.*, § 43.

[30] *Ibid.*, § 114.

[31] *Ibid.*, § 241.

[32] Peter L. Berger – Thomas Luckmann, *The Social Construction of Reality. A Treatise in the Sociology of Knowledge* (1966), New York: Anchor Books, 1967, p. 3.

There is no satisfying theoretical formula which can comprehend what people know as reality. That's why the sociology of knowledge has to focus on what people think of as reality in their everyday life. The authors' central question is: "How is it possible that human activity should produce a world of things? In other words, an adequate understanding of the reality 'sui generis' of society requires an inquiry into the manner in which this reality is constructed."[33]

Berger and Luckmann apply the theory of objectification, i.e. they consider reality to be the outcome of the activity of human beings. This reality is maintained by the beliefs of the members of a community.

> The reality of everyday life is not only filled with objectivations; it is only possible because of them. I am constantly surrounded by objects that "proclaim" the subjective intentions of my fellowmen, although I may sometimes have difficulty being quite sure just what it is that a particular object is "proclaiming", especially if it was produced by men whom I have not known well or at all in face to face situations.[34]

Language is an objectification with special importance. "[L]anguage marks the co-ordinates of my life in society and fills that life with meaningful objects."[35] And let me quote a longer paragraph:

> Language has its origin in the face-to-face situation, but can be readily detached from it... The detachment of language lies much more basically in its capacity to communicate meanings that are not direct expressions of subjectivity "here and now". It shares this capacity with other sign systems, but its immense variety and complexity make it much more readily detachable from the face-to-face situation than any other... In this way, language is capable of becoming the objective repository of vast accumulations of meaning and experience, which it can then preserve in time and transmit to following generations.[36]

This reality becomes internalized due to some quasi institutions and processes such as typification, roles, legitimation, and institutionalization. Language has a crucial importance in all these processes, as well as in the case of the past, personal autobiography, and the integration of other realities. This symbolical universe, according to the authors, is in close connection with human nature whose history is actually an externalizing process. Human beings create their reality along this process of externalization.

Berger and Luckmann regard reality as the outcome of everyday activity which is the inevitable consequence of human nature. This reality is essential for being able to gain orientation in a community.

---

[33] *Ibid.*, p. 18.

[34] *Ibid.*, p. 35.

[35] *Ibid.*, p. 22.

[36] *Ibid.*, p. 37.

Reality from a sociological point of view seemed to be a social construction. From the perspective of cognitive evolution, reality seemed to be a construction in a twofold sense: on the one hand our information from the external world gained by perceptions needs considerable completion. (Dawkins[37] speaks about our software of virtual reality. This software provides the sufficient completion to sensory data assuring a manageable background of our everyday activities.) And on the other hand, socially governed representational patterns also play a very important role in our apprehension of reality.

Communications technologies are decisive with regard to representation, and accordingly, to representational patterns. Reality is considered in accordance with the possibilities provided by the given medium of expression: when the formulation of ideas was restricted to verbal and textual expression, reality was considered being out of the reach of everyday experience. When everyday practice becomes accessible and transmittable due to new inventions of communications technology, reality is regarded as a result of communication and human activity, i.e. reality is considered as a construct ruled by communal activity.

---

[37] Richard Dawkins, *Unweaving the Rainbow. Science, Delusion and the Appetite for Wonder*, Houghton Mifflin Company, 1998.

Social Makeup and Solitary Reason*

Friendship is a curious phenomenon. Everyone has a certain sense of it, but defining its exact meaning is no simple task. Similarly to the way Wittgenstein quotes Augustine deliberating on the concept of time, we can say friendship is "[S]omething that we know when no one asks us, but no longer know when we are supposed to give an account of it".[1] I do not mention Wittgenstein incidenttally. He was sceptical of attempts to arrive at a final meaning, theoretical analysis if you like – in some cases quite explicitly. Wittgenstein calls attention to difficulties that arise when we think of language as Socrates understood it in the *Theaetetus*, i.e. when we consider "speech" as if it was "the composition of names". Distinguishing between composite and simple elements[2], essential and inessential features of an object[3], as well as the existence and non-existence of the object of a proposition,[4] all create unsolvable anomalies. These anomalies can be eliminated only by adopting a basically different conception of language.[5]

Considering friendship, we find something similar in Plato's *Lysis*. After the cumbersome deliberation of different aspects of friendship, Socrates concludes:

> Well, Lysis and Menexenos, we have made ourselves rather ridiculous today, I, an old man, and you children. For our hearers here will carry away the report that though we conceive ourselves to be friends with each other … we have not as yet been able to discover what we mean by a friend.[6]

In first paragraph of the present chapter I mentioned Plato and Wittgenstein,[7] two crucial figures at significant turning points in the history of phi-

---

* First published in Laczkó Sándor and Dékány András (eds.), *Lábjegyzetek Platónhoz 4. A barátság* [Footnotes to Plato, vol. 4: Friendship], Szeged: Pro Philosophia Szegediensi Alapítvány Librarius, 2005, pp. 317-330.

[1] Ludwig Wittgenstein, *Philosophical Investigations*, transl. by G. E. M. Anscombe, Oxford: Basil Blackwell, 1963, § 89.

[2] *Ibid.*, § 47.

[3] *Ibid.*, § 62.

[4] *Ibid.*, § 79.

[5] The same basic conviction of Wittgenstein is of course present in those remarks which deal with the function or limits of philosophy. See *ibid.*, § 123, § 124, and § 125.

[6] Plato, "Lysis", transl. by J. Wright, in: E. Hamilton and H. Cairns (eds.), *The Collected Dialogues of Plato Including the Letters*, Princeton, New Jersey: Princeton University Press, 1989, p. 168 (223b).

[7] Plato is obviously considered to be the philosopher at the time of the transition from primary orality to literacy. Wittgenstein's philosophy was "directly influenced by phenomena of a

losophy to indication that I would like to examine friendship not from the aspect of ethics, (though the notion of friendship emerged mostly within the context of ethics), but from the perspective of communications technology, and its impact on the history of philosophy. I consider friendship to be a concept deeply embedded in the culture of literacy, although it does not fit organically into its epistemological, ontological, and metaphysical framework. This ambivalence is quite obvious since literacy is determined by the representational system of alphabetical writing, which was excellently suited to the creation of the intellectual and institutional framework of solitude, as opposed to the representational skills of orality which are bound up with social intercourse. To prove my thesis, I will first roughly outline the concept of friendship, then sum up the main results of the so-called *Toronto School*, followed by an examination of Merlin Donald's cognitive evolutionary survey. In conclusion, I will show that recently, there has developed a concept of friendship that fit felicitously into the primordial human social makeup and, as opposed to the unique and sublime concept of the last twenty-five centuries, friendship tends to be dispersed.

The Concept of Friendship in the Age of Literacy

In Plato's *Symposium*, the wickedness of mankind was the reason for dividing the primordial "one", a whole human nature into two parts. This division ends in the desire and pursuit of the whole, which is called love.[8] Friendship and love are very close to each other, *viz.* love inspires friendship.[9] According to the desire of the whole,

> when this boy lover – or any lover, for that matter – is fortunate enough to meet his other half, they are both so intoxicated with affection, with friendship, and with love, that they cannot bear to let each other out of sight for a single instant. It is such reunions as these that impel men to spend their lives together, although they may be hard put to it to say what they really want with one another, and indeed, they purely sexual pleasures

---

post-literal type". See J. C. Nyíri, "Post-Literacy as a Source of Twentieth-Century Philosophy", *Synthese* 2002/2, pp. 185-199.

[8] "And so all this to-do is a relic of that original state of ours, when we were whole, and now, when we are longing for and following after the primeval wholeness, we say we are in love. For there was a time, I repeat, when we were one, but now, for our sins, God has scattered us abroad..." (Plato, "Symposium", transl. by M. Joyce, in *The Collected Dialogues of Plato*, p. 545 [193a]).

[9] "For I suppose it does not suit the rulers for their subjects to indulge in high thinking, or in staunch *friendship and fellowship, which Love more than anything is likely to beget.*" (*Ibid.*, p. 536 [182c], italics are mine.)

of their friendship could hardly account for the huge delight they take in one another's company.[10]

Aristotle created quite an elaborate concept of friendship. At its core was a sense of commonness, *viz.* a friend can be someone whom we love, can be considered "another self"[11], and "friendship depends on community".[12]

Casting a glance at some prominent thinkers of modern philosophy (I have in mind in particular Francis Bacon and Rene Descartes), we can see that according to their theoretical frameworks, ethics and friendship had gained a variety of meaning.

Bacon's main aim, to renew science on the basis of experience, made it possible for him to notice special characteristics of human intercourse.

For friendship maketh indeed a fair day in the affections, from storm and tempests; but it maketh daylight in the understanding, out of darkness, and confusion of thoughts. Neither is this to be understood only of faithful counsel, which a man receiveth from his friend; but before you come to that, certain it is, that whosoever hath his mind fraught with many thoughts, his wits and understanding do clarify and break up, in the communicating and discoursing with another; he tosseth his thoughts more easily; he marshalleth them more orderly, he seeth how they look when they are turned into words: finally, he waxeth wiser than himself; and that more by an hour's discourse, than by a day's meditation. It was well said by Themistocles, to the king of Persia, *that speech was like cloth of Arras, opened and put abroad; whereby the imagery doth appear in figure; whereas in thoughts they lie but as in packs*. Neither is this second fruit of friendship, in opening the understanding, restrained only to such friends as are able to give a man counsel; (they indeed are best;) but even without that, a man learneth of himself, and bringeth his own thoughts to light, and whetteth his wits as against a stone, which itself cuts not. In a word, a man were better relate himself to a statue, or picture, than to suffer his thoughts to pass in smother.[13]

Descartes' main aim was to re-establish human knowledge, i.e. to create a solid and unquestionable foundation for knowledge and, as opposed to Bacon, a foundation that has nothing to do with empirical data, but which is provided by the intellectual effort of the individual mind. However, Descartes considered ethics a fruit of the tree whose root is metaphysics. He did not find it necessary to create a systematic ethical theory. As he wrote:

---

[10] *Ibid.*, p. 545 (192c).

[11] "[H]e is related to his friend as to himself (for his friend is another self' *Nicomachean Ethics* Book 9/4 (Aristotle's *Nicomachean Ethics*, transl. by W. D. Ross). D. Ross calls attention to 1161b 18, 1161b 28, 1166a 32, 1169b 6, 1170b 7. (David Ross, *Aristotle*, London: Routledge, 1995, p. 237.)

[12] *Nicomachean Ethics*, Book 8/9.

[13] Francis Bacon, "Friendship", in Francis Bacon, *Essays*, London: Aldine Press, 1968, pp. 83 f.

> It is true that normally I refuse to write down my thoughts concerning morality. I have two reasons for this. One is that there is no other subject in which malicious people can so readily find pretexts for vilifying me; and the other is that I believe only sovereigns, or those authorized by them, have the right to concern themselves with regulating the morals of other people.[14]

However, his main thrust did not require a theory that takes human intercourse into consideration. His so-called "provisional moral code" in *Discourse on the Method* is devoted to the avoidance of mistakes, and to the attainment of happiness, to some extent until certainty can be established. In his *Meditations,* it becomes clear that to know what is true and good makes us free and unable to make a mistake.[15] Actually, as Ernst Gellner in his posthumous book *Language and Solitude* noted:

> He [Descartes] thought it would be possible to judge the culture in which he had been reared from the vantage point of a solitary individual purified by doubt, who accepts nothing other than that which his own reason compels him to accept. Cosmic exile … was, above all cultural exile. It expresses extreme distrust of culture, one's own and all others. Moreover, Descartes felt an acute contempt for culture, which he called 'custom and example' and considered to be the source of all error.[16]

According to Gellner, "The Crusoe tradition, which begins with Descartes, finds its supreme expression in Hume and Kant".[17] On Kant, Ian Hacking has some enlightening words: "What did Kant do? One thing was to make something quite new out of ethics", as Hacking starts his answer.

> In ancient Greece, the topic of ethics had been the good life. Values were out there in the world, and the good life could be perceived and, with diligence, lived. After that, after divine ethics, after Humean naturalized ethics, and after much else, Kant made ethics utterly internal, the private duty of reason.[18]

In his *Metaphysics of Morals*, Kant deliberates friendship in terms of love, respect, and a certain kind of union. As he put it:

---

[14] *The Philosophical Writings of Descartes*, 3 vols., transl. by John Cottingham, Robert Stoothoff, Dugald Murdoch, and Anthony Kenny, Cambridge: Cambridge University Press, 1988, vol. 3, p. 326.

[15] "if I always saw clearly what was true and good, I should never have to deliberate about the right judgement or choice; in that case, although I should be wholly free, it would be impossible for me ever to be in a state of indifference" (*The Philosophical Writings of Descartes,* vol. 2, p. 40).

[16] Ernest Gellner, *Language and Solitude. Wittgenstein, Malinowski and the Habsburg Dilemma,* Cambridge: University Press, 1998, p. 43.

[17] *Ibid.*, p. 182.

[18] Ian Hacking, "Self-Improvement", in: Ian Hacking, *Historical Ontology*, Cambridge, MA: Harvard University Press, 2002, p. 119.

> Friendship (considered in its perfection) is the union of two persons through equal mutual love and respect. It is easy to see that this is an ideal of each participating and sharing sympathetically in the other's well-being through the morally good will that unites them...[19]

According to his general theoretical conviction, Kant emphasizes that friendship "is never for a moment safe from interruptions if it is allowed to rest on feelings, and if this mutual sympathy and self-surrender are not subjected to principles or rules preventing excessive familiarity and limiting mutual love by requirements of respect."[20]

Kant speaks about a "moral friendship" as opposed to a pragmatic one, whose special characteristic is rooted in a special freedom to be able to communicate one's ideas, "to reveal himself" on the basis of confidence and shared views of things. The fruit of moral friendship is that "He is not completely alone with his thoughts, as in a prison, but enjoys a freedom he cannot have with the masses, among whom he must shut himself up in himself".[21]

---

[19] Immanuel Kant, *The Metaphysics of Morals*, transl. by Mary Gregor, Cambridge University Press, 1993, p. 261.

[20] "Although it is sweet to feel in such possession of each other as approaches fusion into one person, friendship is something so delicate (*teneritas amicitiae*) that it is never for a moment safe from interruptions if it is allowed to rest on feelings, and if this mutual sympathy and self-surrender are not subjected to principles or rules preventing excessive familiarity and limiting mutual love by requirements of respect. Such interruptions are common among uncultivated people, although they do not always result in a split (for the rabble fight and make up). Such people cannot part with each other, and yet they cannot be at one with each other since they need quarrels in order to savor the sweetness of being united in reconciliation. But any case the love in friendship can not be an affect; for emotion is blind in its choice, and after a while it goes up in smoke." (Kant, *The Metaphysics of Morals*, p. 262.)

[21] "Man is a being meant for society (though he is also an unsociable one), and cultivating the social state he feels strongly the need to *reveal* himself to others (even with no ulterior purpose). But on the other hand, hemmed in and cautioned by fear of misuse others may make of his disclosing his thoughts, he finds himself constrained *to lock up* in himself a good part of his judgments (especially those about other people). He would like to discuss with someone what he thinks about his associates, the government religion and so forth, but he cannot risk it: partly because the other person, while prudently keeping back his own judgments, might use this to harm him, and conceal his own, so that he would lose something of the other's respect by presenting himself quite candidly to him.

If he finds someone intelligent – someone who, moreover, shares his general outlook on things – with whom he need not be anxious about this danger but reveal himself with complete confidence, he can then air his views. He is not completely alone with his thoughts, as in a prison, but enjoys a freedom he cannot have with the masses, among whom he must shut himself up in himself...

This (merely moral friendship) is not just an ideal but (like blacks swans) actually exists here and there in its perfection. But that (pragmatic) friendship, which burdens itself in with the ends of other men, although out of love, can have neither the purity nor the completeness

As we can see, the notion of friendship reflects upon human relationships and even intercourse. Epistemology, however, and the metaphysical framework of the individual – particularly after the Cartesian turn in philosophy – do not contemplate any kind of social embeddedness.

In the Platonic dialogues, the notion of friendship is based on the myth of the primordial unity of divided human beings on the one hand, and on the other hand, theoretical deliberations, i.e. in the conceptual framework of love and creation, more precisely: *poiesis*. Aristotle distinguishes different kinds of friendship.[22] The desirable, i.e. the "[p]erfect friendship is the friendship of men who are good, and alike in virtue; for these wish well alike to each other qua good, and they are good themselves. Now those who wish well to their friends for their sake are most truly friends; for *they do this by reason of own nature* and not incidentally". [23] (Italics mine) That is, reason gains crucial importance regarding friendship. According to Bacon's considerations, friendship is useful

---

requisite for a precisely determinant maxim; it is an ideal of one's wishes, which knows no bounds in its rational concept but which must always be very limited in experience." (*Ibid.*, p. 263.)

[22] Three different kinds of friendship according to Aristotle are friendships based on utility, friendships based on pleasure, and perfect friendship. See Aristotle, *Nicomachean Ethics*, Book 8/2-3.

[23] "Perfect friendship is the friendship of men who are good, and alike in virtue; for these wish well alike to each other qua good, and they are good themselves. Now those who wish well to their friends for their sake are most truly friends; for they do this by reason of own nature and not incidentally; therefore their friendship lasts as long as they are good-and goodness is an enduring thing. And each is good without qualification and to his friend, for the good are both good without qualification and useful to each other. So too they are pleasant; for the good are pleasant both without qualification and to each other, since to each his own activities and others like them are pleasurable, and the actions of the good are the same or like. And such a friendship is as might be expected permanent, since there meet in it all the qualities that friends should have. For all friendship is for the sake of good or of pleasure-good or pleasure either in the abstract or such as will be enjoyed by him who has the friendly feeling-and is based on a certain resemblance; and to a friendship of good men all the qualities we have named belong in virtue of the nature of the friends themselves; for in the case of this kind of friendship the other qualities also are alike in both friends, and that which is good without qualification is also without qualification pleasant, and these are the most lovable qualities. Love and friendship therefore are found most and in their best form between such men.

But it is natural that such friendships should be infrequent; for such men are rare. Further, such friendship requires time and familiarity; as the proverb says, men cannot know each other till they have 'eaten salt together'; nor can they admit each other to friendship or be friends till each has been found lovable and been trusted by each. Those who quickly show the marks of friendship to each other wish to be friends, but are not friends unless they both are lovable and know the fact; for a wish for friendship may arise quickly, but friendship does not." (Aristotle, *Nicomachean Ethics*, Book 8/2-3.)

in everyday activity as well as intellectual reasoning. Kant's notion of friendship is determined – just like his ethics – by the "private duty of reason".

Other people, friends, everyday intercourse, social, and institutional frameworks we meet in everyday practice provided the basis of the need to investigate these phenomena from a philosophical point of view. However, in the intellectual milieu of literacy, it was not easy to integrate the other individual into the general image of the Cartesian ego. That may be the reason for the unique and admirable role of the friend, as opposed to the role of simply the "other fellow" with whom we can have some relation for different practical purposes.

Summing up theoretical considerations regarding friendship, we can say that although practical moments are present, the claim for an abstract and rational foundation to friendship is dominant.[24] All practical aspects of friendship are considered shallow and secondary in comparison to the abstract and generally defined notion of friendship, which is noble, and in its perfection represents higher principles. This kind of abstraction is in full harmony with other topics of philosophy in the age of literacy. However, being locked in the prison of our own thoughts as opposed to being able to reveal ourselves via communication with a friend seems to be in contradiction with the notion of individual solitary reason. What good could a friend be within the framework of the Platonic *ideas*, or for the Cartesian *ego*, or even the Kantian "Transcendental Ego"?[25]

This ambivalence can be understood as the ambivalence of the socially embedded human mind in the age of literacy, i.e. an ambivalence of social make up and solitary reason. Next, I'm going to outline the main characteristics of literacy within the framework of the cognitive evolution of the human mind.

Solitary Reason

How could a pure, representational technology (in this case, alphabetical writing) have an impact upon cognitive customs and intellectual attitudes? How could the purely technical means of writing establish rational reasoning? To answer these questions, I would first like to delineate the main characteristics of verbal expression, and then compare the noetical world of primary oral and literate cultures.

Verbal expression, as a representational framework, automatically requires a certain degree of generality. This becomes quite clear if we compare it to

[24] The only exception seems to be Bacon. His methodological commitment to induction can explain his attitude towards the notion of friendship.

[25] For some interesting reflections on the "transcendental Ego" see Gellner, *Solitude and Culture*, p. 45.

image-like or pictorial representation. The importance of similarities,[26] analyzing-reordering, and the linear structure of verbal expression have a considerable effect on cognitive capacity and the thought process. As alphabetical writing made it possible to detach verbal representation from its multi-dimensional and situational roots (i.e. live intercourse), the inclination towards abstract and general considerations significantly increased.

This change is clearly visible in the investigations of Harold A. Innis, Walter J. Ong, and Eric A. Havelock, just to mention some representatives of the Toronto School. The Ongian notions of orality and literacy are based on studies which analyzed early written records from the new perspective of communications technology. These investigations uncovered characteristic patterns of preservation, the interweaving of ideas, cognitive capacity, and even the very organisation of communities.

In Ong's terms, in a primary oral culture there is no way to preserve and pass down knowledge and experiences beyond the realm of live intercourse. Accordingly, accumulation and renewal are slow and hindered processes. With the invention of alphabetical writing, incorporating new ideas or even the criticism of old ones become possible. Although written records exist to preserve knowledge, the possibility of reviewing, comparing, and reconsidering the heritage of the past gives birth to criticism. On the other hand, new demands emerge as regards vocabulary for the purpose of being able to express general and eternal truths.[27]

Havelock, regarding Heraclitus' recorded heritage, called attention to the fact that

> [o]ut of a total of some one hundred and thirty sayings, no less than forty-four, or thirty-four per cent, are preoccupied with the necessity *to find a new and better language*, or a new and more correct mode of experience, or are obsessed with the rejection of current methods of communication and current experience.[28] (Italics mine.)

---

[26] "All that words can deal with, however, are similarities. The simple reason for all this is that words, with the exception of proper names, relation words, and syntactical devices, are mere conventional symbols for similarities. Although differences are just as perceptible as similarities," words are not able to cope with them. "But that these differences are not statable in words does not mean that they are ineffable, for they are clearly communicable in non-verbal ways." W. M. Ivins, Jr., *Prints and Visual Communication*, Cambridge, Massachusetts: Harvard University Press 1953, p. 139.

[27] For more details on the differences between the oral and literate mind see Walter J. Ong, *Orality and Literacy: The Technologizing of the Word*, London: Methuen, 1982, pp. 36-46; Havelock, *The Greek Concept of Justice. From its Shadow in Homer to Its Substance in Plato,* Cambridge, Massachusetts and London: Harvard Univ. Press, 1978. In the present volume see pp. 16 f.

[28] "Heraclitus, no less than his audience, is compelled to have daily acquaintance with Hesiod's world. Yet it is precisely this acquaintance that he would wish to disrupt. He does

However, as time went on, difficulties arising from the changing requirements of the changing representational framework emerged in the form of different theoretical considerations, puzzles, and in terminology: reflection upon these changes became more and more implicit. Plato tried to define abstract notions of intelligence, courage, temperance, justice, and friendship, etc.;[29] Aristotle aimed to improve and control our thought-process with the help of syllogism.

In a primary oral culture, knowledge of various facts, norms, and rules was available only through live intercourse. As knowledge became attainable through written records as well as via communication with others, the question of reality and the accessibility of knowledge about reality became central issues. Accordingly, the question of true knowledge and its foundation gained importance.

The main questions of medieval metaphysics were also related to quite general issues. Medieval thought about universals and the question of transcendency and immanency with regard to the foundation of being gave rise to the central questions of metaphysics. Considerations connected to these questions had an abstract and complicated conceptual background. The concepts used to grasp reality created a specific hierarchical world far from everyday experience.

Metaphysics in the Modern Age turned to the question of the reliability of knowledge, particularly knowledge that originates from the past or from the senses. This specific distrust can be connected to the perception of a change in the technology for disseminating ideas – the technology of the printing press. Due the unreliability of manuscripts (and of diagrams, illustrations, and pictorial

---

not want to live in this world. No wonder, then, that he is obsessed by the difficulty of making statements which shall be from his point of view correct: and once a statement has been correctly worded, it appears no less difficult for an audience to take in what has been said or to communicate it to others. Out of a total of some one hundred and thirty sayings, no less than forty-four, or thirty-four percent, are preoccupied with the necessity to find a new and better language, or a new and more correct mode of experience, or are obsessed with the rejection of current methods of communication and current experience. This statistic is striking in a man who in later tradition was represented as a philosopher of materialism and fiery flux. Fire, in fact, is mentioned in only five of this [*sic*] sayings and has been inserted in the text of three more by scholars who perhaps have been a little overzealous to justify the traditional estimate of him. Clearly, if we take his *ipsissima verba* seriously, his preoccupation with problems of vocabulary, and the psychological response to vocabulary, must be regarded as central." (Eric A. Havelock, *The Literate Revolution in Greece and Its Cultural Consequences*, Princeton University Press, 1982, pp. 220-260, http://nyitottegyetem.phil-inst.hu/kmfil/kmkt/hav_pre3.htm.)

[29] Cf. "Plato is aware that he is engaged in a process of 'naming names', fixing them, we might say, as new names, new insofar as they are to become symbols of conceptual identities." Eric A. Havelock, *The Greek Concept of Justice*, p. 327.

statements as well) became conspicuous. The most important task for metaphysics, then, was to find the correct manner of cognition, which meant to simultaneously create a firm basis for the sciences. The connection between modern science and the technology of the printing press is quite obvious.[30] And let us note that it was the technology of printing texts in thousands of identical copies that gave rise to the important and novel concepts of originality and authorship.

By the 18th century, literacy had reached its zenith. "Thinking in alphabetical characters" (to use the formula of the Hungarian scholar István Hajnal) was the obvious intellectual milieu of the time, and the printing press had a deep effect on scientific development. Reliance on science, and the importance of the scientific method were stronger than ever before. Kant's Copernican Turn is thus seen to be built upon two postulates: the insufficiency of logic, and the necessity of synthetic *a priori* statements, which then created possibilities for mathematics and sciences.[31] These postulates mirror the ambivalence of the state of literacy: a simultaneous disappointment and trust in the results of literacy.

Despite this ambivalence, reaching beyond the individual mind (i.e. to discover its social-cultural and physical embeddedness) was a hopeless enterprise for quite a long time. Regarding the inability to reach beyond the "immanence of mind", significant changes took place in the 20th century. By this time, communication had become equipped with the technology of photography, film, and the "electrical speech machine", i.e. the telephone. If we have a look at the metaphysics of the 20th century, we see that the glory of logic and the scientific method was no longer apparent. As new equipments for communication directly mediate numerous moments of everyday life, metaphysics discovered everyday practice and tried to integrate it into its framework. Heidegger took a decisive step when he left the immanence of the mind and tried to focus on everyday life. He emphasized our "thrown-ness" into the world as opposed to the traditional dualism of object and subject. The world, earlier considered an object to be known and separate from the human intellect, emerged as a unity of references helping our orientation. Similarly, Wittgenstein focused his attention on embeddedness in everyday practice, in opposition to the theoretical analyzing attitude of earlier metaphysicians. As emphasized in his *Philosophical Investigations,* the authentic approach to language is given through its use, not through an abstract logical method. Likewise, Heidegger's attempt to reveal the primordial meaning of *logos* – a renewal of the ancient

30 See e.g. Elisabeth L. Eisenstein, *The Printing Press as an Agent of Change: Communications and Cultural Transformation in Early-Modern Europe* I-II. Cambridge: Cambridge University Press, 1979.

31 For more details, see the chapter "The Iconic Turn in Metaphysics" above, p. 100.

concept of language as a kind of creation – was aimed at setting limits on the validity of modern logic.

Social Makeup

Due to the technology of the 20th century, philosophy was able to rediscover the importance of our social makeup in a two-fold sense. As I mentioned, thanks to developments in communications technologies, particularities could be transmitted with the same ease as similarities and generalities. Thus, everyday practice emerged on the horizon of theoretical considerations. The results of recent scientific research on cognition are very useful, even inevitable, to rethinking some main issues.

Merlin Donald, in his book *Origins of the Modern Mind,* outlined the history of cognitive evolution. According to his theory, the development of representational skills was in close relation to social intelligence.[32] Considering social intelligence, Donald emphasizes that "it is clear that language was the final step, and that presymbolic forms of social intelligence must have been its foundation."[33] Donald distinguishes three main transitions in human cognitive evolution: the shifts from episodic to mimetic, then to mythical, and finally to theoretical culture consisted of new representational skills. Episodic memory, complete with mimetic representation, could create community, with special habits and organization, providing a certain kind of identity. Verbal representation enabled the birth of mythical constructions as institutions of orientation. The birth of theoretical culture was bound up with an effective *external symbolic storing system.*[34]

This apparently technological invention established considerable alterations in cognitive habits.

> [T]heoretic culture broke with the metaphoric style of meaning in oral-mythic culture. Where narrative and myth attribute *significances,* theory is not concerned with significance in the same sense at all. Rather than modeling events by infusing them with meaning and linking them by analogy, theory dissects, analyzes, states laws and formulas, establishes principles and taxonomies, and determines procedures for the verification and analysis of information. It depends for its advanced development on specialized memory devices, languages, and grammars.[35]

[32] Regarding social intelligence, Donald relies on the investigations of Robin I. M. Dunbar. Merlin Donald, *Origins of the Modern Mind. Three Stages in the Evolution of Culture and Cognition,* Cambridge, MA: Harvard Univ. Press, 1991, pp. 10 and 137.

[33] Merlin Donald, *Origins of the Modern Mind,* p. 137

[34] *Ibid.*, p. 320 (in the present volume see p. 78).

[35] *Ibid.*, pp. 274 f.

Donald's reconstruction of cognitive evolution, and the argumentation for it clearly reveal "that the evolution of cognitive structure *at the modular level* might have continued well beyond the point at which physical evolution had stopped." Cultural patterns have an unexpectedly far-reaching effect on cognition, even "in terms of its fundamental neurological organization".[36]

His theory suggests that our representational skills are determined by what I call 'social makeup' (i.e. our sociability), and inventions related to communication have a decisive impact on intellectual and social institutions.

Friendship Re-visited

Considering friendship from a pedestrian point of view, intimacy and trust seem to be two of its crucial, decisive elements. These two qualities can be considered necessary and sufficient conditions of friendship even within the framework of perfect or moral friendship. However, these two attributes have significance, to variable degrees, in everyday intercourse as well. Clear-cut alphabetical reason honoured ideal friendship for its integrity regarding metaphysical presupposetions, even though some discrepancies arose when considering practical, experience-like deliberations and speculative constructions.

Friendship did not appear in the philosophy of Wittgenstein, or even Heidegger. However, they are two crucial representatives of the attempt to leave the speculative-analytical attitude for the sake of the integrity of everyday experience and the theoretical-philosophical processing of it.

Considering recent notions of friendship, one can see that its honoured and unique image has started to become dispersed: it appears that everyone with whom we have an active relationship that includes trust and intimacy with more than average intensity, we now consider to be a friend. According to my thesis, this change is quite obvious from the perspective of communications technology. That is, since we have the opportunity to communicate easily at any time, and nearly everywhere, the space for solitary reasoning decreases. Since the institutional framework of communication has radically changed, we must re-structure our everyday activities, scheduling, and, accordingly, the ideals of human relationships.

Let me quote from an article in *The Guardian*:

> There are powerful reasons why we should create these bonds, even if we only start when we are older. The phenomenon of later births means families take up a smaller percentage of our lives. We wait years to have children, and we could be 70 before we become grandparents for the first time. We have more time available, and fewer familial responsibilities, than the generations before us. We all want to feel needed and valued by others. It is possible for friends to fill that need, but only if we work at it.

[36] *Ibid.*, p. 14.

> It isn't easy, because friendship is a subtle dance, and no one wants to be explicitly pursued when it's unwelcome, or explicitly dropped when they are not wanted. Nor does it come with any guarantees. People are unpredictable. But we need to play the game of friendship.[37]

During the centuries when institutions of social intercourse were determined by solitary reason, friendship as a residual of our primordial social make-up was an elaborated topic, however poorly integrated into the metaphysical frame of reference. Recently, scientific and some philosophical theories about human nature and reasoning emphasize the role of sociability, and technological means made possible a return to (or at least a similarity to), communicational patterns which were dominant before the age of literacy. Due to new technological tools for keeping in touch in a more natural (multimodal) manner, an almost face-to-face communication conquers the literate mind in its everyday practice. Thus we communicate more, and with increased intensity. We can alter intensive connections in time for content or purpose, and at the same time, we are capable of keeping in rather intimate touch from afar. Accordingly, it is conceivable to gain and lose friends without losing one's integrity. That is, the more the ease keeping in touch (and we may add, the more obvious the importance of sociability is), the less conspicuous the role of friendship is.

[37] "What are friends for?" in: *The Guardian*, January 24, 2005.

Being Mobile: Cognitive Multiplicity*

Being mobile is crucial for the organization of our everyday work and life. New communications technologies provide the opportunity to organize activities more and more flexibly and effectively. However, this increasing ease of organizational activities has its special price: the traditional division among the different fields of activities – work and spare time, official engagement and privacy – and concomitant scheduling has started to brake down. Mobile phones have often been considered devices of disturbance. At the same time, with the collapse of the traditional framework of everyday work and life, cellular phones are becoming indispensable devices to manage exactly these activities.

This apparent contradiction is easily dissolvable by making its roots and workings visible. According to the position I will side with in this paper, the means used to express and disseminate ideas are in close inter-relation with our cognitive capacities. The rapid adaptation of cellular phones indicates that some newly needed cognitive capacities are actually in our possession, ready to be exploited. I will here rely on Merlin Donald's cognitive-evolutionary work on the one hand, and the conclusions of the so-called Toronto School on the other. I would like to propose that aversions to new instruments are rooted in the habits and institutions created by devices of communications accustomed to earlier. First, I will point to some main characteristics of mobile usage, from the specific perspective of communications technologies. I will then analyze the changes indicated above from a cognitive evolutionary perspective. By way of conclusion, I will consider attempts at modelling cognition utilizing the conceptual means of communications technology.

Using Mobiles – from the Historical Perspective of Communications Technologies

Cellular phones allow us to communicate instantly and comfortably across vast distances. They offer different discreet ways to contact colleagues and friends. Mobile telephony provides the possibility of "perpetual presence"; we can be reached for whatever purpose and we can keep in touch with others regardless of their local conditions and engagement. This kind of freedom provides a feeling of security: one is always able to ask for help. On the one hand, there arise the

* First published in Kristóf Nyíri (ed.), *Mobile Understanding. The Epistemology of Ubiquitous Communication*, Vienna: Passagen Verlag, 2006, pp. 79-90.

difficulties caused by perpetual multitasking, in particular the switch between various physical, psychological, and computational conditions.[1]

The quick adaptation to and the spread of the cellular phone do not mirror these difficulties. Deliberating on the reason for them, we encounter institutions that have developed during the age of literacy. Describing these institutions, I will first compare the communicational patterns of orality with that of literacy, and then go on to sketch some special characteristics of verbal expression.

In the age of primary orality the only possibility to maintain and preserve knowledge was communicating; that is, sustaining living language – a *storage language* as Havelock puts it – which was "devised orally for the purpose of survival."[2] Such restriction required a special technology to weave ideas together, and to transmit awareness of the new facts of life. According to this special technology the mode of expression was additive, redundant, and the expressions and words used were very closely embedded in concrete situations. Intercourse was empathetic, participatory and agonistic.[3]

Literacy, due to the invention of alphabetical writing as a new kind of storage system, with its new kinds of limits[4], created new habits in cognition compared to those of orality. Alphabetical writing reinforced the dominance of verbal expression and created a special one-sidedness in thinking. Verbal face to face communication is situational and multimedial, it does not have to be complete in its verbal aspects. Ever since verbally expressed knowledge became recordable in writing, a new way of composing ideas became necessary to compensate for the absence of the living situation. This necessity induced thinking to retire from everyday life: there emerged general subjects, abstract concepts detached from the human life-world, linearly structured arguments, and the linear notion of time.[5] The new means of expressing ideas made it possible to

---

[1] Cf. Joshua S. Rubinstein, David E. Meyer and Jeffrey E. Evans, "Executive Control of Cognitive Processes in Task Switching", in: *Journal of Experimental Psychology - Human Perception and Performance*, vol. 27, no. 4 (2001). The conclusion of the study is that multitasking may actually take more time in the end, i.e. the outcome regarding efficiency is questionable. Let us not however that, considering the potentials of cognitive evolution and the plasticity of the human brain, the idea of an adaptation to new circumstances is not at all far-fetched.

[2] E. A. Havelock *The Muse Learns to Write. Reflections on Orality and Literacy from Antiquity to the Present*, New Haven, CT: Yale University Press, 1986, p. 59.

[3] See Walter J. Ong, *Orality and Literacy: The Technologizing of the Word*, London: Methuen, 1982, pp. 36-46.

[4] The holistic perception of situations which is given in live intercourse, here has to be constructed with only the help of written words. The situation, gestures, and mimics have to be substituted by words.

[5] See Ong, *Orality and Literacy*, especially pp. 31-57, 103-112, and regarding linearity p. 143; E. A. Havelock, *The Greek Concept of Justice: From its Shadow in Homer to Its Substance in Plato*, Cambridge, MA: Harvard University Press, 1978, E. A. Havelock, *The Muse Learns to*

record general and comprehensive statements. The written text became a reliable and effective medium of preserving knowledge. But let us note that during the centuries of literacy, for a long time to come, the same did not hold for the dimension of pictorial representation.

Considering the representational frameworks of verbal versus pictorial expression, we can see that verbal representation demands a certain re-structuring of any phenomenon. In the case of pictorial representations, we can capture the whole as well as the details of an image to gain a holistic view of the given phenomenon. By contrast, words can represent a given phenomenon only re-structured: general features, circumstances, details, modalities, etc. have to be described as being ordered according to certain priorities. In live intercourse, the particular situation furnishes some components or parts of the information, but in the case of verbal expression, there is a certain degree of generality. Moreover, as William M. Ivins, Jr. put it: "All that words can deal with ... are similarities. The simple reason for all this is that words, with the exception of proper names, relation words, and syntactical devices, are mere conventional symbols for similarities." Differences, Ivins suggests, are just as perceptible as similarities; however, words are not able to cope with them.[6]

This structure of verbal representation creates a certain distance from the phenomena by adapting to the possibilities and limitations created by the use of words. Being able to represent something via words is tantamount to being able to think according to structured priorities. The use of words, i.e. verbal representation in a given situation, requires a special point of view, a special perspec-

---

Write; and Peter Gendolla, "Punktzeit. Zur Erfahrung in der Informationsgesellschaft" in: R. Wendorf (ed.), *Im Netz der Zeit*, Stuttgart: Hirzel, 1989.

[6] "But that these differences are not statable in words does not mean that they are ineffable, for they are clearly communicable in non-verbal ways." (William M. Ivins, Jr., *Prints and Visual Communication*, Cambridge, MA: Harvard University Press, 1953, p. 139.) Ivins offers considerations about Western thinking as the outcome of a special technological inability: "I have a notion that much of the philosophical theory of the past can eventually be traced back to the fact, that whereas it was possible after a fashion to describe or define objects by the use of arbitrary and exactly repeatable word symbols addressed, mediately or immediately, to the ear, it was not possible to describe or define them by exactly repeatable images addressed to the eye." (*ibid.*, p. 62) Cf. also Merlin Donald' emphasis on differences between pictorial and verbal representation. This suggests that pictorial representation might be more direct and less mediated than verbal representation. Donald differentiates three kinds of visual symbolic innovation: pictorial, ideographic, and phonological writing. The main difference between them is that "writing generally does not depend upon perceptual isomorphisms and metaphors, whereas the pictorial modes does. The pictorial mode is primarily dependent upon visual norms and does not necessarily engage linguistic structures" (Merlin Donald, *Origins of the Modern Mind. Three Stages in the Evolution of Culture and Cognition*, Cambridge, MA: Harvard University Press, 1991, p. 278), and words "must have been inherently general-purpose constructs from the start" (*ibid.* p. 252).

tive, to use the terminology of Michael Tomasello.[7] Accordingly, in everyday practice intentions and perspectives will dominate verbal expression, as opposed to the holistic grasp of the world which is mirrored by pictorial representation.

This tendency became stronger due to the invention of external storage systems whose most effective version was phonetic writing. However, replacing the live situation with mute words required considerable cognitive and intellectual effort and caused difficulties. Epistemology as formed during the centuries of fully developed literacy is determined by the dualism of the knowing subject opposed to the object to be known. Moreover, the subject is considered to be an individual mind which can gain knowledge about the external world by its own cognitive efforts and/or private experiences.[8] Written texts and the concomitant activities created the basis for the conviction that it is systematic individual effort which can yield true value. Still, language, the main tool to express ideas, became an object of criticism because it often proved insufficient for the purpose of rational argumentation, exact description and regulation. Language was increasingly considered to be the means of rational thinking which is to be improved.

The 20th century was a "language-textured" age[9], which is not really surprising. Since grasping reality and language are hardly separable as long as the main representational framework is provided by language, metaphysics and the philosophy of language overlapped. This overlap became very conspicuous by the 20th century. Surprisingly, or perhaps not so surprisingly, that century, in its later decades, came to labelled as the age of "secondary orality". During this period several inventions occurred in communications technologies: photography, film, gramophone, radio were introduced into everyday life. As the details and elements of day-to-day activities became easily communicable, philosophical reflection, which hitherto had been bound up with texts, turned to phenomena not previously transmittable. With particular, everyday, details of life

---

[7] See Michael Tomasello, *The Cultural Origins of Human Cognition*, Cambridge, MA: Harvard University Press, 1999.

[8] Cf. Ernest Gellner: "there is atomistic individualism, which sees the individual building his cognitive world (and indeed any other) by orderly, step-by-step, individual effort, possibly maintaining cooperative relationships with others similarly engaged, but without this fundamentally affecting the nature of the enterprise, which in the end is solitary." (Ernest Gellner, *Language and Solitude: Wittgenstein, Malinowski and the Habsburg Dilemma*, Cambridge: Cambridge University Press, 1998, p. 181.)

[9] Cf. e.g. George F. Sefler, *Language and the World: A Methodological Synthesis Within the Writings of Martin Heidegger and Ludwig Wittgenstein*, Atlantic Highlands, NJ: Humanities Press, 1974, p. 195. See also Ernest Gellner's observations on the so-called linguistic turn, that is, how language became the focus of philosophy (E. Gellner, "The Crisis in the Humanities and the Mainstream of Philosophy", in J. H. Plumb (ed.), *Crisis in the Humanities*, Harmondworth, Middlesex: Penguin Books, 1964).

appearing on the horizon of philosophical meditations, unexpected anomalies occurred within the texture of traditional philosophy.

The development described above can be considered as a change in the framework of cognitive and social activities. Accordingly, at first, individuals communal and communicative, then atomistic and rational, and recently, re-integrated into communities, were order of day. In what follows, I will now bring into the picture the cognitive evolutionary theory of Merlin Donald.

Morals from Cognitive Evolution

Donald summed up the change in cognitive skills during the last two millennia as a development leading to "a new human representational apparatus". As he writes: "The entire theoretic complex of visual symbolic devices, memory management skills, and related metalinguistic skills was completely novel".[10] The steps towards this new representational apparatus presupposes the existence of certain "external devices" which

> allow an iterative, interactive thought process to operate repetitively on its own products; and, more importantly, the thought process itself can be largely externalized and institutionalized. Since external memory devices can provide a linkage to all preceding levels of cognitive evolution, they also serve in the construction of an integrative field, where the products of various types of thinking can be juxtaposed and combined.[11]

The development of external memory[12] was a long-term evolution of representational skills which were in close relation to social intelligence.[13] Donald emphasizes that "it is clear that language was the final step, and that presymbolic forms of social intelligence must have been its foundation".[14] Donald dis-

---

[10] Donald, *Origins of the Modern Mind*, p. 358.

[11] *Ibid.*

[12] I would like to emphasize that verbal and pictorial representations from a cognitive-evolutionary point of view seem to bear different functions. Although both of them have special importance in the maintenance of social cohesion, it was pictorial representation that first provided the opportunity to bridge the gap between, on the one hand, internally processed knowledge and experience, and on the other, a kind of external storage system. Donald supposes that pictorial representations emerged in the framework of oral-mythical culture and he considers pictorial representation as the first step towards an external storage system. As he puts it: "the first pictorial images themselves were also external representations. They existed outside of the individual, rather than in visual memory. Therefore a technological bridge was under construction that would eventually connect the biological individual with an external memory." (*Ibid.*, pp. 284)

[13] Regarding social intelligence Donald relies on the investigations of Robin I. M. Dunbar. See Donald, *Origins of the Modern Mind*, pp. 10 and 137.

[14] *Ibid.*, p. 137.

tinguishes three main transitions in human cognitive evolution: a shift from episodic to mimetic, then to mythical, and finally to theoretical culture. Each of these changes means the emergence of a new kind of representation as well as increasing load on biological memory[15]. Episodic memory supplemented with mimetic representation created community with special habits and organization, conferring a sense of identity on the members of community. Verbal representation built upon the mimetic one was the first such stage when mythical constructions playing the role of orienting world-views and structures emerged. The transition from mythical to theoretical culture presupposes the existence of an effective external storage system.

> [T]he first two evolutionary transitions would have greatly increased the load on biological memory. However, the final step in this tremendous cognitive expansion might have *reduced* the load on some aspects of biological memory, by gradually shifting many storage tasks onto the newly developed E[xternal]S[ymbolic]S[toring system]. At the very least, the existence of the ESS must have forced a great change in priorities and memory organization.[16]

Donald's reconstruction of cognitive evolution and the argumentation for it clearly reveal

> that the evolution of cognitive structure *at the modular level* might have continued well beyond the point at which physical evolution had stopped. Cultures restructure the mind, not only in terms of its specific contents, which are obviously culture-bound, but also in terms of its fundamental neurological organization. Whether the organization is vested in a parallel set of specific brain adaptations or not (and obviously at times it is not), the brain sets fewer constraints than formerly thought on the process of cognitive evolution. Culture can literally reconfigure the use patterns of the brain; and it is probably a safe inference from our current knowledge of cerebral plasticity that those patterns of use determine much about how the exceptionally plastic human central nervous system is ultimately organized, in terms of cognitive structure.[17]

Donald realizes the difficulties regarding the cognitive function of external memory. Reckoning with the fact that "our view of humanity is alongside the

---

[15] "Human memory had, from its inception, expanded the range of primate memory. The earliest form of hominid culture, mimetic culture, depended on an expansion in the self-representational systems of the brain and created the initial base for semantic memory storage, which consisted initially of representational action scenarios reflected in mime, gesture, craft, and skill. With the evolution of speech and narrative ability, there were even greater increases in the load on biological memory, adding not only the storage networks for phonological rules and the lexicon in its entirety but also a very large store of narrative conceptual knowledge." (*Ibid.*, p. 319.)

[16] *Ibid.*, p. 320.

[17] *Ibid.*, p. 14.

machinery of our mind",[18] he tries to make its function clear with the help of metaphors. He uses the metaphor of computer R[andom]A[ccess]M[emory] which is supplemented by an external memory for the purpose of expanding the computer's capabilities. Donald stresses that though the capability of RAM can be increased, it has physical limits just as in the case of biological memory. The other metaphor he uses is also from the realm of computer science. As he puts it:

> External memory is best defined in functional terms: it is the *exact* external analogue of internal, or biological memory, namely, a storage and retrieval system that allows humans to accumulate experience and knowledge. We do not possess any ready theoretical frameworks in psychology from which to view external memory. Fortunately, there is an excellent point of comparison in the field of computing science: networks.[19]

Networks can be seen as the society of computers. Connectedness provides the possibility that computers do not have to retain the same "cognitive" /computational abilities. They can use "the resources of the system ... shared, and the system functions as a unit larger than any of its individual components."[20] Of course the emergence a functioning network presupposes definite structural changes as well. Being equipped with an interface is necessary for the use of external memory and external capabilities. However "once plugged in [humans'] skills are determined by both the network and their own biological inheritance".[21]

The cognitive evolutionary point of view sheds light on the modifications of cognitive structure. It provides solid background for the cultural-historical observations of the scholar. The development of social organization and regulation and, last but not least, the history of philosophy can be easily considered to be attempts at solving difficulties raised by the limitations of the dominant representational framework of the time.[22]

Modeling Cognition

Attempts to describe human understanding have deep roots in Western metaphysics and epistemology. Recent investigations in different fields like psychology, artificial intelligence, neurobiology and cognitive science reveal that

---

[18] *Ibid.*, p. 382.

[19] *Ibid.*, p. 309.

[20] *Ibid.*, p. 310.

[21] *Ibid.*, p. 311.

[22] Think of the dominance of the general over the particular due to the dominance of verbal expression. Think also, of the dilemma of precision vs. richness of an expression, reflecting the role of tacit coefficient knowledge in language as well as in scientific rationality. On these issues see above, "Changing Media – A Perennial Challenge for Philosophy", p. 45.

the traditional metaphysical model of the human mind is mistaken. Let us note that contemporary scientific research aiming at modeling cognition merits philosophical considerations not only in its achievements, but also in its failures. The consequences and force of the traditional notion of dualism – *res extensa* and *res cogitans* – and the limitations of the literate and rational mind can be highlighted by reflecting upon insufficiencies emerging in the framework of cognitive science.

One main branch of cognitive science is classical cognitivism, to use the term by Andy Clark.[23] Classical cognitivism considers the mind to be a calculating and symbol-manipulating machinery, i.e. it approaches the mind in the spirit of Cartesian dualism and focuses on abilities which are amenable through language. The other main branch of cognitive science is connectionism which is built on the model of the neural network of the brain. This is clearly a useful approach, but the question of choice regarding the necessary input remains rather artificial, since connectionist models can best work in the framework of restricted, well-designed, micro-worlds.

The tendency to focus on only restricted areas of the external world is an obvious consequence of the dualistic approach. A similar, earlier consequence was the Cartesian ego, isolated, and supposed to follow its systematically constructed set of rules. The dualism of the knowing subject and the object to be known suggests that the connection between the two poles is one-sided. The subject wants to gain information about the world as it is in order to gain true knowledge and to find general rules of nature. Moreover, this subject is considered to be individual, general in its functioning, and independent of its circumstances. Accordingly, modeling cognition easily tends to be restricted to a special problem, and of course cognition is easily identifiable with symbol-manipulation driven by certain algorithms, i.e. a kind of calculation. Andy Clark points out some other difficulties arising from the traditional approach. As he puts it:

> In fact, the methodology of rational reconstruction can mislead in several crucial ways. First, the immediate replacement of real physical quantities with symbolic items can obscure opportunistic strategies that involve acting upon or otherwise exploiting the real world as an aid to problem solving. (Recall the 007 principle.) Second, conceptualizing the problem in terms of an input-output mapping likewise invites a view of cognition as *passive computation*. That is, it depicts the output phase as the rehearsal of problem solution.[24]

[23] Andy Clark, *Microcognition. Philosophy, Cognitive Science, and Parallel Distributed Processing*, Cambridge, MA: MIT Press, 1989, ch. 1.1.1.

[24] Andy Clark, *Being There. Putting Brain, Body, and World Together Again*, Cambridge, MA: MIT Press, 1997, p. 81. Regarding the "007 principle" see *ibid.*, p. 46.

Considering cognition as symbol-manipulation is only apparently obvious. Language is the main forum of formulating and disseminating ideas. Due to the institutions developed during the high tide of literacy, and, in particular, due to the philosophical delusions arising from the former, language is considered to be insufficient for the purpose of rational, i.e. clear-cut and exact thinking: additional means of logic seem to be necessary in order to improve the workings of language. From the perspective of writing – which is actually manipulation of arbitrary signs while serving the technical means of rationality[25] – and that of logic, symbol-manipulation seems to be the only effective way by which to describe cognition. However, the results of developmental psychology and evolutionary surveys suggest quite a different concept of cognition.

Andy Clark calls attention to the fact that the gradually evolved structure of different biological and cognitive functions is hardly understandable from a teleological perspective. He further refers to certain achievements of developmental psychology, and, finally, to cases which reveal how the external world provides an orientational potential. With the help of this array of references Clark shows that cognition is not isolated from its environment and indeed biologically, socially, and culturally embedded. Clark's famous concept of the *extended mind* is built upon the expedient scaffolding which can be found in the functioning of the simplest as well as the most complicated biological organisms.

> We build 'designer environments' in which human reason is able to far outstrip the computational ambit of the unaugmented biological brain. Advanced reason is thus above all the realm of the *scaffolded* brain: the brain in its bodily context, interacting with a complex world of physical and social structures.[26]

Although Clark refers to Donald's theory,[27] he does not exploit its argumentative potential. The reconstruction of the evolution of representational abilities is, I believe, convincing enough not to question the importance of external structures and institutions.[28] Modifications in the way humans handle

---

[25] Cf. István Hajnal's views on literacy in *Technika, művelődés: Tanulmányok*, Budapest: História, 1993, p. 18, quoted above in "Objectification or Spiritual Involvement?", pp. 5 f.

[26] Clark, *Being There*, p. 191.

[27] *Ibid.*, pp. 206 and 210. See also Andy Clark, *Natural-Born Cyborgs: Minds, Technologies, and the Future of Human Intelligence*, Oxford: Oxford University Press, 2003, p. 79.

[28] Consider e.g. the following passage: "The regions of the brain that are most characteristically human – especially the great expanses of the frontal and anterior temporal lobes – are likely to be the most malleable neurological structures in nature, taking on many forms. They are configurable and reconfigurable to a remarkable degree, because their recourses are allocated on a *competitive* basis to the many input paths impinging on them. In effect, *the physical structure of mind has become less and less fixed as neocortical evolution has progressed.* This leaves room not only for the kinds of radical reconfiguration introduced by

and grasp their world as it is revealed from the perspective of communications technologies suggest accepting external institutions as important organizers of cognition.

Conclusion

As Clark points out: "Mind cannot be usefully extended willy-nilly into the world... But the overall picture is of a rather special kind of user/artifact relationship – one in which the artifact is reliably present, frequently used, personally "tailored", and deeply trusted."[29]

Just like in the case of mobile phones. Indeed, cellular phones are conducive to the above conditions. Judging by the widespread use of mobile telephony, one can hardly question that it is becoming an important part of everyday life. The use of mobile phones caused perceptible modifications in the structure of everyday activities as it became an essential element of the institutional framework of communication. These modifications and the quick adaptation reveal ancient capacities which were concealed by the institutions of literacy. Currently we create external structures which require exactly those patterns of behaviour which were buried by the institutions that emerged as answers to the demands of literacy. Shifting the emphasis let me quote Kristóf Nyíri: a "mobile phone is ... a machine which corresponds to deep, primordial human communicational urges".[30]

Finally, I would like to focus briefly on the work of István Hajnal, the Hungarian scholar whose basic contribution was a *theory of objectification.* Hajnal points out that humans tend to create a structure of "scaffolding", both material and institutional. This is the process of objectification; it causes experiences, emotions, and skills to become visible, manifest. To make them manifest means to detach them from the natural ground they originate from. Hence they can serve as a support for, as well as a limitation of, our activities. He claims that from the perspective of the social sciences, the process of objectification renders the general structure of a community or society and the particularities of its every day life manageable in a single coherent theoretical framework.

---

literacy but also (presumably) for larger differences between the brains of individual human beings than are possible in primates. It also leaves room for further cognitive restructuring, possibly in fundamental ways." (Donald, *Origins of the Modern Mind*, pp. 380 f.)

[29] Clark, *Being There*, p. 217.

[30] Kristóf Nyíri, "Introduction: From the Information Society to Knowledge Communities", in: Kristóf Nyíri (ed.), *Mobile Communication. Essays on Cognition and Community*, Vienna: Passagen Verlag, 2003, p. 12.

Hajnal's concept of literacy defines a special culture which becomes possible due to the technical-practical invention of alphabetical writing and leads to ways of thinking and cognition never before possible. The invention of a new recording system of alphabetical writing has created the demand for novel ways of thinking and, accordingly, new cognitive skills. The mere technical means of writing was able to affect human intelligence so deeply simply because it was a special kind of objectification, a kind of material support for the thought process.

Clark's emphasis is quite similar even though he looks at the matter from a very different point of view. He refers to an important distinction in Donald's theory. Donald differentiates mythic from theoretic culture and, as Clark stresses, the necessary conditions of the latter were not simply symbolic invention, but "[it was] the *process* of externally encoded cognitive change and discovery".[31] Clark speaks about "objectification" with regard to language. As he puts it: "The cultural tool of public language gives us not just labels but whole, structured, recursive *systems* for the encoding, objectification, and communication of thoughts and ideas."[32]

If one considering Hajnal's comprehensive theory (especially his concept of literacy), Donald's reconstruction of cognitive evolution, and the Toronto School's findings as to the impact of various technological means of communication on everyday activity, as well as on intellectual disposition and cognitive capacity, Clark's theory is not really ground-breaking. The three former theories attempt to clarify the close inter-relatedness of the individual with its environment, while Clark takes this idea further by focusing on the fact that the individual cannot be detached from its (technical-technological, institutional, and cultural) environment. The concept of the extended mind is in harmony with some main tendencies occurring in twentieth-century philosophy (recall Heidegger, recall Wittgenstein) and it has scientific support as well. This is clearly a significant step towards a holistic grasp of cognition, opposed to a dualistic one. Epistemology seems well on its way to leave the immanence of the individual mind.

[31] Clark, *Natural-Born Cyborgs*, p. 79, quoting Donald, *Origins of the Modern Mind*, p. 343.
[32] *Ibid.*, p. 72.

Solitude, Culture, and the Technology of Communication*

Philosophical Investigations was first published in 1953. It is, of course, widely regarded as the most important posthumous publication of Wittgenstein, containing the main elements of his later philosophy.[1] William M. Ivins', Jr. *Prints and Visual Communication* was published in the same year. The main topics of these two books are quite different, but we can find critiques of Plato's philosophy in both. In the following section, I would like to compare these critical considerations not only with each other, but also against the background of Ernest Gellner's notion of *abstract universalistic individualism* and the rival position which he calls *romantic communalism*.

I will first recount, very briefly, Wittgenstein's criticism of Plato in the *Philosophical Investigations*, and then describe Gellner's historically-oriented point of view of the above-mentioned traditions in philosophy. Finally, I shall cite arguments by Ivins concerning Plato which – I hope to show – shed new light on Gellner's criticism.

Wittgenstein quotes these sentences from the *Theaetetus*:

> If I make no mistake, I have heard some people say this: there is no definition of the primary elements – so to speak – out of which we and everything else are composed; for everything that exists in its own right can only be *named*, no other determination is possible, neither that it *is* nor that it *is not*... But what exists in its own right has to be named ... without any other determination. In consequence it is impossible to give an account of any primary element; for it, nothing is possible but the bare name; its name is all it has. But just as what consists of these primary elements is itself complex, so the names of the elements become descriptive language by being compounded together. For the essence of speech is the composition of names.[2]

Wittgenstein then calls attention to difficulties which arise when we think of language along the lines Socrates understood it in the *Theaetetus*, i.e. when we consider "speech" as if it was "the composition of names". Distinguishing between composite and simple elements[3], essential and inessential features of an

---

* First published in R. Haller and K. Puhl (eds.), *Wittgenstein and the Future of Philosophy. A Reassessment after 50 Years*, vol. 1, Kirchberg am Wechsel: ALWS, 2001, pp. 428-434.

[1] Ernest Gellner, *Language and Solitude. Wittgenstein, Malinowski and the Habsburg Dilemma*, Cambridge: University Press 1998, p. 71.

[2] Ludwig Wittgenstein, *Philosophical Investigations*, transl. by G. E. M. Anscombe, Oxford: Basil Blackwell, 1963, § 46.

[3] *Ibid.*, § 47.

object[4], as well as the existence and non-existence of the object of a proposition,[5] create unsolvable anomalies.

> When I say "N is dead", then something like the following may hold for the meaning of the name "N": I believe that a human being has lived, whom I (1) have seen in such-and-such places, who (2) looked like this (picture), (3) has done such-and-such things, and (4) bore the name "N" in social life. – Asked what I understand by "N", I should enumerate all or some of these points, and different ones on different occasions. So my definition of "N" would perhaps be "the man of whom all this is true". – But if some point now proves false? – Shall I be prepared to declare the proposition "N is dead" false – even if it is only something which strikes me as incidental that has turned out false? But where are the bound of incidental? – If I had given a definition of the name in such a case, I should now be ready to alter it.[6]

This longer quote clearly shows that there are conspicuous difficulties when one tries to describe/prescribe the truth conditions of a statement by exactly-defined elements and by the rules of logic. Wittgenstein emphasizes that these difficulties arise only because language is considered out of the context of its usage. That is, these anomalies can be eliminated only by adopting a basically different perspective on language. This perspective is put forward in the later philosophy of Wittgenstein – as opposed to that of the *Tractatus.* According to the *Philosophical Investigations*, the only authentic approach to language is given through its use – not through distinctions and concepts formulated with the help of some abstract logic.[7] Traditional concepts of logic such as sentence, word, and symbol are defined from a new perspective, in a wider context, i.e. in the light of linguistic usage.

> [T]he meaning of a word is its use in the language.[8]
> Our language can be seen as an ancient city: a maze of little streets and squares, of old and new houses, and of houses with additions from various periods; and this surrounded by a multitude of new boroughs with straight regular streets and uniform houses.[9]

Language is not a static phenomenon, it changes in the course of its use. (Just like definitions in accordance with the context of their usage.) "[T]he speaking of language is part of an activity, or a form of life." The term "lan-

[4] *Ibid.*, § 62.
[5] *Ibid.*, § 79.
[6] *Ibid.*
[7] Wittgenstein's same basic conviction is, of course, present in those remarks which deal with the function or limits of philosophy. See *Philosophical Investigations*, §123, §124, and §125.
[8] *Ibid.*, § 43.
[9] *Ibid.*, §18. The regularity and uniformity of new boroughs can be considered a reference to the tradition of – in Gellner's terminology – abstract universalistic individualism.

guage-game" underscores this fact.[10] The so-called "private language argument"[11] emphasizes the communal character of language, i.e. the fact that speaking a language means obeying rules[12] – agreements in definitions and implicit presuppositions[13] which gain sense only in a community.

The very idea that language is a form of life[14] implicitly entails that language is closely bound to culture. But Wittgenstein is not concerned with culture directly. His main interest relates to philosophical questions such as meaning, understanding, sentence, concepts of logic, foundations of mathematics, sensory data, and the opposition of idealism and realism.[15]

Gellner unambiguously places Wittgenstein's later philosophy in the tradition of *romantic communalism*, contrasting it with the *Tractatus*, which he sees as representing the tradition of *abstract universalistic individualism*.[16] By this contrasting of philosophical traditions, Gellner is able to relate Wittgenstein's philosophy to a given cultural-political situation. According to Gellner's interpretation, Wittgenstein tries to find a way out from the solitude into which *abstract individualism* has led. He doesn't succeed; Gellner believes that his pursuit of the way out leads to another kind of solitude. The pursuit of Wittgenstein mirrors the intellectual and political atmosphere of Kakania, i.e. the Habsburg Empire.

> ...what does make sense is that, having worked out one philosophy in terms of abstract and universal reason and having found that it led to intolerable solitude, he replaced it by another philosophy, inspired by the one thing which, in the pervasive atmosphere of Kakania, was widely recognised to be the alternative to reason: namely roots. Roots, not

---

[10] *Ibid.*, § 23. See also: "I shall call the whole, consisting of language and actions into which it is woven, the language-game." (*Ibid.*, § 7.)

[11] On the private language argument from the point of view of the technology of communication see J. C. Nyíri, *Tradition and Individuality*, Dordrecht, Boston, London: Kluwer Academic Publishers 1992, pp. 105-113.

[12] Wittgenstein, *Philosophical Investigations*, § 202.

[13] "If language is to be a means of communication there must be agreement not only in definitions but also … in judgements." (*Ibid.*, § 242.)

[14] *Ibid.*, § 19.

[15] TS 225: i-iv as quoted by Nyíri (Kristóf Nyíri, *Ludwig Wittgenstein*, Budapest: Kossuth Könyvkiadó, 1983, p. 83).

[16] "The *Tractatus* had portrayed the human condition, cognitively, morally, semantically – in the spirit of an uncompromising, extreme, virtually comic universalism. Not only science but the mystical as well were described as ineluctably identical in all men. Culture, religion, gender, none of these, or any other specific characterisation, make any difference to anything that really matters in our life. We are all alike, the differences belong to a zone of insignificant. – And just these differences lay at the heart of Wittgenstein's later philosophy." (Gellner, *Language and Solitude*, p. 105.)

logical form, dictate our speech and confer our identity on us and limit our world. He renamed it 'a form of life'.[17]

Abstract individualism, that is the Crusoe tradition which, according to Gellner, started with Descartes, sees human knowledge as developing step by step due to the intellectual effort of the individual. It doesn't take into account either communal or institutional circumstances as regards cognition. Human beings gain knowledge by their own cognitive efforts and/or private experiences. Acquisition of knowledge is possible only through abstract, distinct, and well-defined concepts, in the course of a systematic process. By contrast, *romantic communalism* stresses the pervasive power of tradition, culture, and community. This tradition has a much older history than the other, but became systematized only later.

> Once upon a time mankind lived, by and large, in closed intimate communities, governed by practices simultaneously geared both to maintaining internal order and adjusting to nature ... The criteria adapted for judging the acceptability of practices ... were, so to speak, self-validating, traditional. They were not systematised; no attempt was to deduce them either from some supposedly self-evident general premise or from some single authoritative revelation ... Then, one day, a new style of cognition emerged, which separated the referential inquiry into nature from the concern with internal social harmony and which, by means which are still only partly understood, succeeded in acquiring astonishingly accurate, general and consensus-securing understanding of the environment.[18]

Gellner stresses that both traditions have their own function, i.e. they have explanatory force as regards different segments of our life: the tradition of *abstract individualism* helps to explain, e.g., the development of science and the functioning of the economy; and the tradition of *romantic communalism* can help, e.g., to find a way to harmonize individual, communal, and even natural aspects, both factual and normative. Gellner criticized Wittgenstein because of a special one-sidedness: "He lived out the two options in reverse order, and decreed that the romantic one constituted normality, and the universalistic one was a disease of language. He projected his own rather bizarre development onto the history of thought."[19]

Criticizing Wittgenstein, Gellner believes: "The real intellectual problems that modern society faces consist, in very large part, of the relationship between the two styles."[20] In what follows, I would like to show that these "two styles"

---

[17] *Ibid.*, p. 106.
[18] *Ibid.*, p. 189.
[19] *Ibid.*, p. 191.
[20] *Ibid.*, p. 190.

do not necessarily exclude each other as regards their explanatory power; moreover, in some respects they can complement each other.

In Ivins' *Prints and Visual Communication*, he focuses on the history of rediscovering images, and the main causes of their oblivion. Ivins calls attention to the hindering technical factor behind the fact that the expressive potential of images has been forgotten. This technical incapability means that it was not possible to duplicate images precisely, without distortions. Later, after the invention of printed pictures, due to the different techniques of engraving, images remained rather improper records of information.[21] Accordingly, verbal expression (both in its oral and written forms) could also dominate pictures in the future.

This fact has far-reaching consequence with regards to the main elements of literate considerations.

> Plato's Ideas and Aristotle's forms, essences, and definitions, are specimens of this transference of reality from the object to the exactly repeatable and therefore seemingly permanent verbal formula. An essence, in fact, is not part of the object but part of the definition. Also, I believe, the well-known notions of substance and attributable qualities can be derived from this operational dependence upon exactly repeatable verbal descriptions and definitions – for the very linear order in which words have to be used results in a syntactical time order analysis of qualities that actually are simultaneous and so intermingled and interrelated that no quality can be removed from one of the bundles of qualities we call objects without changing both it and all the other qualities.[22]

Accordingly, ancient Greeks, and even "thinking people" of the 19th century, lack the proper technology with which they can take into account, and handle, particularities; so they think in generalities.[23]

Ivins explicates a deeper cause of the limits of verbal expression.

> As a matter of fact, the moment that anyone seriously tries to describe an object carefully and accurately in word, his attempt takes the form of an interminably long and prolix rigmarole that few persons have either the patience or the intelligence to understand. A serious attempt to describe even the most simple piece of machinery, such, let us say, as a kitchen can-opener…
> As we think about this it becomes obvious why the tool-maker wants not a written description which he is called upon to make but a series of carefully made drawings accompanied by terse specifications of materials, dimensions, and, especially, of

---

[21] Ivins calls attention to another important aspect of graphic representation when speaking about the "scheme for laying lines" that became "grammars and syntaxes which, while making handmade pictorial statements possible, also greatly restricted and influenced their power of statement." (William M. Ivins, Jr., *Prints and Visual Communication*, Cambridge, Massachusetts: Harvard University Press, 1953, p. 164.)

[22] *Ibid.*, p. 63.

[23] *Ibid.*, p. 91.

tolerances. A purely verbal description demands of the tool-maker that he takes a series of arbitrary, abstract, vague, word symbols arranged in a linear order and translate them into concrete forms of material in a three-dimensional space in which there is no linear or time order and in which everything exists simultaneously.[24]

The dominance of verbal expression requires considerable abstraction from the living context. This implies that we need various cognitive or technical instruments to be able to articulate and communicate ideas.[25] ***Abstract universalistic individualism*** can be considered as a tradition which fabricates cognitive instruments to be able to handle these capacities, and the lack of other capacities. ***Romantic communalism*** as a tradition represents the other side – being, that is, less precise in concepts and closer to the "immense wealth of living shapes". The idea that communications technology has great importance as regards the cognitive styles of human beings is made plausible by the considerations sketched above, and can be documented by evidence drawn from the history of culture and philosophy.[26]

Considering Wittgenstein's and Ivins' criticisms of Plato, one can notice that although their point of view is totally different (thus their criticisms can be complementary to each other) they talk about the same phenomenon. That is, Ivins stresses those technical and intellectual characteristics of verbal expression (as opposed to the pictorial) which have a considerable orientational effect on the thought process (and which, we might add, become more radical in the age of literacy); and Wittgenstein emphasizes the anomalies which arise from the limited capability of the evolving literate mind. The previously quoted paragraphs clearly show that language (even by Plato) is considered in accordance with generalities, beyond any kind of particularity.

---

[24] *Ibid.*, pp. 57 f. Cf. Michael Polanyi's ideas as regards the effect that in the course of reproducing or describing reality, we have to choose between either being precise or being close to the richness of living context. It is impossible to simultaneously grasp precision and the "immense wealth of living shapes", the richness of the whole context. See M. Polanyi, *Personal Knowledge - Towards a Post-Critical Philosophy,* London: Routledge & Kegan Paul, 1962, p. 86 f. (In the present volume see p. 45.)

[25] One of Gellner's earlier studies touches upon this problem in the course of explicating the function of philosophy. See Ernest Gellner, "The Crisis in the Humanities and the Mainstream of Philosophy", in J. H. Plumb (ed.), *Crisis in the Humanities*, Middlesex: Penguin Books, 1964 p. 68.

[26] See Elisabeth Eisenstein, *The Printing Press as an Agent of Change: Communications and Cultural Transformations in Early-Modern Europe,* Cambridge: University Press, 1979; E. A. Havelock, *The Greek Concept of Justice. From its Shadow in Homer to Its Substance in Plato,* Cambridge, Mass.: Harvard University Press, 1978; E. A. Havelock, *The Muse Learns to Write. Reflections on Orality and Literacy from Antiquity to the Present*, New Haven: Yale University Press, 1986; and Ivins, *Prints and Visual Communication.*

Finally, I'd like to show the importance of pictures/images in Wittgenstein's argumentation, which provides further similarities between Ivins' and Wittgenstein's considerations. When Wittgenstein considers simple and composite elements or the nature of definition,[27] he uses pictures to clarify his arguments. As he put it: "The *philosophical* question: "Is the visual image of this tree composite, and what are its component parts?" the correct answer is: "That depends on what you understand by 'composite'." (And that is of course not an answer but a rejection of the question.)"[28] That is, images are capable of providing circumstances so directly that there is no need for explanatory details and definitions. As Ivins formulates it, "verbal description demands of the toolmaker that he takes a series of arbitrary, abstract, vague, word symbols arranged in a linear order and translate them into concrete forms."

The long centuries of literacy conceal the special features of language and other possible ways of expression. This provides the opportunity for strengthening the positions of *abstract universalistic individualism* which limited the intellect to the world of generalities and exact definitions. Due to evolving communications technology, particularities (such as the aspects of communal everyday life) are also accessible to abstract reasoning.

[27] I have in mind Wittgenstein's example of "N is dead" where one of his points regards what N looks like, that is the picture of him which means a regressus ad infinitum, since the description of his look requires further definitions. See the quotation on p. 86 above.

[28] Wittgenstein, *Philosophical Investigations*, § 47.

IMAGES AND WORDS

The Iconic Turn in Metaphysics*

Images gained importance in communication in the 19th century due to the technology of photography, and subsequently that of film. As Koskinen puts it in a study on the future of picture messaging: "The Internet has created new roles for visual expression and graphics. The image has become a tool. At the same time, it has become a central part of the interface between technology and its users."[1] Recently, widespread mobile communication has been supplemented with the possibility of sending multimedia messages (MMS). Despite the exciting option of sending and receiving sound and images as easily as short text messages, the use of MMS is less pervasive than had been expected. An obvious question arises: what is the reason for this? Considering the gradual rise of the importance of pictures, it is surprising that in mobile communication they fail to play the expected role. According to my thesis, the reasons for this fact can be found in the special character of mobile telephony on the one hand, and of literacy on the other. Mobile communication provides an excellent possibility for keeping in touch with colleagues, family members, and friends. SMS messages – brief messages in the form of texts – are convenient for sharing information in cases when the message has to be silent, unnoticed by others, and not disturbing ongoing face-to-face communication. MMS messages can fulfill these requirements but they cannot be sufficiently informative, because we are not equipped with shared and obviously interpretable schemes of images. Due to this insufficiency, there are only a very few people who are able to understand our *ad hoc* pictorial statements precisely, if at all, with the trivial consequence that MMS cannot be as effective as SMS. The reason for this resides in our cultural heritage, which is rooted in literacy, leading our minds to express ideas primarily in words.

We have gradually lost the ability to apply and use the informative force and effectiveness of images. This process can clearly be traced through the history of philosophy, more precisely in that of metaphysics. Though philosophy, and metaphysics probably even more, is generally bound up with literacy, it might be greatly enlightening to cast a glance upon a very different kind of metaphysics – *viz.* the metaphysics of pictorial statements, which can be found in ancient Egyptian thought. I would like to compare the metaphysics of ancient Egypt formulated through pictures to Western metaphysics expressed in words. I

* First published in Kristóf Nyíri (ed.), *A Sense of Place. The Global and the Local in Mobile Communication*, Vienna: Passagen Verlag, 2005, pp. 395-403.

[1] Ilpo Koskinen, Esko Kurvinen, Turo-Kimmo Lehtonen, *Mobile Image*, Helsinki: Edita Publishing, 2002, p. 13.

believe this comparison can, ultimately, help explain the surprising fact that MMS is not as popular as it was expected to be.

How can such an abstract discipline that is apparently far removed from everyday practice be relevant to a practical issue like the use of MMS in everyday communication? In order to answer this question, first I would like to make clear what I mean when I use the word "metaphysics". Then, to prove my thesis, I shall briefly characterize the two abovementioned kinds of metaphysics. By way of conclusion, I will summarize the difference between them and at the same time I will attempt to show that communicating via pictures is still a world to re-conquer, a challenge to our literate mind.

What Does It Mean: Metaphysics?

From an etymological point of view we can speak about metaphysics since the time of the first editorial ordering of the Aristotelian corpus in the 1st century B.C. – "Meta ta physika" was the designation given to the work that came next after *Physics*.

According to its definition, metaphysics can be thought of "first philosophy", and the foundation of any other scientific investigation. According to this initial definition, associated with the name of Aristotle, of what was later called metaphysics, it is "a science which investigates being as being and the attributes which belong to this in virtue of its own nature".[2]

As the contemporary scholar, Jorge J. E. Gracia puts it in his book *Metaphysics and Its Tasks*: "metaphysics tries to identify and relate the most general categories and to establish the relations of less general categories to the most general ones." This is why "metaphysics is fundamental – it is logically prior to every other discipline – and inescapable – metaphysical views are presupposed by every other view. It is for this reason that one can regard metaphysics as the categorial foundation of all knowledge."[3] To formulate it in a less esoteric way, metaphysical presuppositions provide a framework for each theoretical consideration which determines the nature and context of the experienced or investigated phenomena. Without this background, orientation would be impossible. In this sense metaphysics is inevitable. "As Kant put it:

[2] "There is a science which investigates being as being and the attributes which belong to this in virtue of its own nature. Now this is not the same as any of the so-called special sciences; for none of these others treats universally of being as being. They cut off a part of being and investigate the attribute of this part; this is what the mathematical sciences for instance do. Now since we are seeking the first principles and the highest causes, clearly there must be some thing to which these belong in virtue of its own nature." (Aristotle, *Metaphysics*, transl. by W. D. Ross, Book IV, 1.)

[3] Jorge J. E. Gracia, *Metaphysics and Its Task: The Search for the Categorial Foundation of Knowledge*, Albany, NY: SUNY Press 1999, p. xviii.

'That the human spirit will ever give up metaphysical speculations is as little to be expected as that we should prefer to give up breathing altogether, in order to avoid impure air.'"[4]

The Holism of the Iconic Mind

If we consider these special features of metaphysics, it is not an overstatement to speak of a metaphysics of ancient Egypt. I will now sketch some special features of Egyptian thought. In doing this, I rely considerably on a book by the Hungarian scholar Márton A. Farkas. I would like to emphasize those characteristics of the ancient Egyptian intellectual-cultural heritage which are in close connection with the special way of formulating ideas on the one hand, and which are strange, sometimes contradictory, to the Western way of thinking. I shall not touch upon the Egyptian recording system as a whole, except for the part of it which determined the intellectual-philosophical background of Egyptian cult and culture.

One special, and for Western thinking quite strange, feature of Egyptian thought is its "complementary" logic. This is in a close relationship with symbolic expression. Anything can be something else at the same time. In pictorial representation, this makes it possible to emphasize different aspects of the given phenomena and allows for the use hieroglyphs as pictures and pictures as hieroglyphs. From the viewpoint of logic, this kind of pluralism means that opposite characteristics complement rather than eliminate each other: *tertium datur*.

Pictorial signs and images used in pictorial writing provide continuous connection to their roots – be it an abstract, a religious, or just a narrative statement. Images can display direct and metaphorical meanings as well as everyday objects and concepts at the same time. The icon of a scroll as determinative or a taxogram was used to indicate different phenomena on the basis of the kind of abstractness. Accordingly, applying the icon of the scroll after the given signs resulted in abstract concepts such as *knowledge* and *truth*, or verbs such as *to know*, *to begin*, etc.[5]

It is in the technology of representation that an implicit theory of reality can be recognized. The Egyptian way of representation makes explicit a merged point of view which delivers all the important moments of the given phenomena, and not just the visible features. This is necessary, because only in such a way

---

[4] Quoted by Gracia, *op. cit.*, p. xiii.

[5] The icon of a house was applied in cases of a kind of closed space, e.g. nest, horizon, nether world, etc. See Attila Márton Farkas, *Filozófia előtti filozófia: Szimbolikus gondolkodás az ókori Egyiptomban* [Philosophy Before Philosophy: Symbolic Thinking in Ancient Egypt], Budapest: Typotex, 2003, p. 144.

can the represented phenomenon and its image be identical. The difference between modern scientific and Egyptian thinking emerges with sharp outlines when we consider the perspective which was thought of in modernity as the art of picturing objects-as-they-are, i.e. in the objective, hence true, manner.

This special way of representation is in close connection with the conviction that to understand pictorial statements is at the same time to create reality. The flexible and plural logic of picturing is complemented with flexible and plural interpretation. To understand a pictorial statement requires the full knowledge of and belief in Egyptian religion, mythology, science and everyday practice, i.e. it needs total intellectual involvement. As pictorial statements can show a wide range of perspectives, they can be interpreted according to the given situation. Moreover, to read, i.e. to understand a picture, is at the same time to be *in* the picture. Just as looking at the pictures of a tomb is at the same time being in the nether world, to know what a picture means is to be in the "virtual" world which is represented by the picture. Since phenomena are mainly pictures and pictures are writing, phenomena are readable pictures. Accordingly, to read a picture is to be in the world which is visualized/reproduced by the picture.

This holistic picture of the universe is in full harmony with the given way of expressing ideas, i.e. pictorial writing. Hieroglyphs and the way they were used supported this holistic view. There is no room here to detail the quasi-syntax of this type of writing, but images can represent practical and general features as well as a narrative background at the same time. Similarly, the phenomena of the universe have practical and general aspects and are related to religious narratives. Due to the sun rising, things come to life: as they become visible, they emerge from the primordial chaos. Accordingly, after sunset, things provisionally die.[6] The sun represents life and the act of creation as much as it represents light. And sunset simultaneously means the revival of the nether, the hidden, invisible world which contains and preserves the invisible, inactive, past and becoming, as well potential instead of actual beings.[7]

Clear-Cut Alphabetical Reason and Changing Media

In ancient Egyptian thinking, reality was bound to language – more precisely to words – and forms, i.e. to images. According to the Memphite theology, in the process of creation, Ptah, who is the God of craftsmen as well, created the world via words and forms, a world which made it possible to know and to understand the physical-natural world, created by Atum, since the world actually could not exist without words and forms, or at worst it could exist in itself as an

[6] *Ibid.*, p. 266.
[7] *Ibid.*, p. 251.

undifferentiated mass.[8] Since words can become manifest in forms, words and forms are inseparable. Accordingly, meaning and name as well as meaning and sign are the same, i.e. reality.[9]

In Western thinking, this unity has disintegrated. The change is obvious in the light of the difference between pictorial and alphabetical writing: The former represents a whole culture and bears creational force at the same time. The latter just reconstitutes the acoustic phenomena of speech in space, but it does not have the creative potential that language was supposed to have. Accordingly, sign and meaning can not be the same, and reality, though its concepts were formulated in texts, was no longer considered something that can be created by written records.

Due to this disruption of the unity of reality and its cognitive-cultural processing, the central issue of metaphysics became first the foundation of reality, then the clear, abstract, and true description of it. Along this path of metaphysics, a special dualism and more and more sophisticated and complicated terminology can be noticed. The dualism appeared in different kinds of oppositions: the inner vs. outer world, immanency vs. transcendency, subject vs. object, etc. The terminology has been changing according to the increasingly complicated solutions of the emerging anomalies on the one hand, and the increasing abstractness supported by alphabetical writing on the other. Alphabetical writing in itself is unable to preserve the connection between live experience and written content. It uses abstract signs of sounds, but it is not able to refer to the actual source of the written word. Since writing was very effective in preserving, accumulating, and reproducing ideas, it was able to become the main forum for objectifying the latter. Though writing is a clear and unambiguous way of recording, there was a price to pay for its effectiveness: all elements of the idea or narrative must be expressed by words. There is no supporting situation/environment or unity of references which can help understanding. Every statement must be understandable in itself with the only help being the context.

As a new kind of storage system with its built-in bounds, alphabetical writing created new habits in cognition compared to orality: general subjects, abstract concepts detached from the human life-world, linearly structured arguments, and the time-scale emerged. The metaphysical mind closes itself up in the world of texts and tries to solve puzzles whose roots are in this closing up, but implicitly reflects upon the new inventions of communication technologies which could directly affect, even if in the long run only, the literate mind.[10]

---

[8] *Ibid.*, p. 281.

[9] *Ibid.*, p. 266.

[10] Cf. Merlin Donald, *Origins of the Modern Mind: Three Stages in the Evolution of Culture and Cognition*, Cambridge, MA: Harvard University Press, 1991.

Plato, who is considered the philosopher of the transition from primary orality to literacy, distinguished between our apprehension of reality and appearance as being knowledge or opinion, respectively. The former is the realm of ideas and the latter is the territory of shadows. These two spheres are not totally separated: the act of remembering provides the bridge between them.[11]

Along the path of metaphysics up to the age of Enlightenment, a dualistic way of grasping reality resulted in an unbridgeable chasm between mind and body. Two different metaphysical traditions – empiricism and rationalism – took shape, which provided the starting point for Kant at the zenith of literacy.

Kant argued that earlier metaphysics was unable to fulfill its real task – which was, in Kant's opinion, to analyze experience[12] – because it possessed only logical tools whose validity did not reach beyond the realm of the mind's conclusions. (Think of the principle of *tertium non datur*. If we apply this to physical phenomena, e.g. opposed forces with the same strength, it results not in a contradiction, but a kind of equilibrium.) Accordingly, logic with its traditional means, *viz.* syllogisms, is not able to capture the system of reality, i.e. nothing can be said about experience. For Kant, this "first sharp dualism", as Cassirer referred to it, and the importance of the validity of mathematics (which is synthetic and *a priori*), took form in the dualism of phenomenal and noumenal worlds. The latter is inaccessible because all of our sense data gain meaning for us owing to a "mental" process whose result is the phenomenal world. This is the limit of our knowledge; "things in themselves" are inaccessible and not knowable.

By the 18th century, literacy had reached its zenith. "Thinking in alphabetical characters"– using the formula of the Hungarian scholar István Hajnal – was the obvious intellectual milieu of the time, and the printing press had a deep effect on scientific development. This new technology helped the accumulation of knowledge, allowing for exact duplication which gave rise to a certain distrust in the traditions of the past on the one hand, and enabled the exact duplication of images and diagrams on the other hand.

The ancient notion of *logos*, long regarded as referring to creative potential, increasingly came to mean logic, i.e. a system of rules which describes the functioning and the proper use of language, and considerable efforts were made to create a faultless and universal language. The reliance on science and the importance of the scientific method were stronger than ever before. As we saw, Kant's Copernican turn can be built upon two postulates: the insufficiency of logic, and the necessity of synthetic *a priori* statements which create the possibility for mathematics and sciences. These postulates mirror the ambi-

---

[11] See the myth of Lethe in Plato's *Republic*, 614b–621b, esp. 621a–b.

[12] Ernest Cassirer, *Kants Leben und Lehre*, Berlin: Bruno Cassirer, 1921, p. 78.

valence of the state of literacy: disappointment and trust in the results of literacy at the same time.

The technology of printing can be considered as a major step towards the use of images in communication, because it afforded for the first time the possibility to use images as carriers of definite meaning, i.e. to duplicate them exactly.[13] Earlier, since images could not be usefully disseminated without exact duplication, they gradually disappeared from effective means of communication and lost their informative force. Thus after the invention of the printing press and especially after the invention of photography, images as it were had to be rediscovered.

Micheal Polanyi speaks about the dilemma of being precise or being close to the "immense wealth of living shapes".[14] William M. Ivins, Jr. sheds light on this: "All that words can deal with ... are similarities. The simple reason for all this is that words, with the exception of proper names, relation words, and syntactical devices, are mere conventional symbols for similarities. Although differences are just as perceptible as similarities, the inability of words to cope with them has given rise to the notion held by many self-consciously hard-headed persons that talk about art is merely an attempt to deal with the ineffable, a thing that for them is completely laughable. But that these differences are not statable in words does not mean that they are ineffable, for they are clearly communicable in non-verbal ways."[15]

By the 20th century, communication was augmented by the technologies of photography, film, and the "electrical speech machine", i.e. the telephone. If we have a look at the metaphysics of the 20th century, we realize that the glory of logic and the scientific method is no longer taken for granted. As new communications equipment directly mediates numerous moments of everyday life, metaphysics discovers everyday practice and tries to integrate it into its framework. Heidegger took a decisive step when he left the immanence of the mind and tried to focus on everyday life. He emphasized our "thrown-ness" into the world as opposed to the traditional dualism of object and subject. The world, which earlier was considered as an object to be known and separate from the human intellect, emerged as the unity of references helping our orientation. Similarly, Wittgenstein focused his attention on embeddedness in everyday

---

[13] Eisenstein emphasized the importance of the "printed image" with regard to scientific development. See Elisabeth L. Eisenstein, *The Printing Press as an Agent of Change: Communications and Cultural Transformation in Early-Modern Europe*, vols. I–II, Cambridge: Cambridge University Press, 1979.

[14] M. Polanyi, *Personal Knowledge - Towards a Post-Critical Philosophy*, London: Routledge & Kegan Paul, 1962, pp. 86 f.

[15] W. M. Ivins, Jr., Prints and Visual Communication, Cambridge, MA: Harvard University Press, 1953, p. 139.

practice as against the theoretical-analytical attitude of earlier metaphysics. As emphasized in the *Philosophical Investigations,* the authentic approach to language is given through its use as opposed to through an abstract logical method. Likewise, Heidegger's attempt to reveal the original meaning of *logos* – a renewal of the ancient concept of language as a kind of creation – was aimed at setting limits on the validity of modern logic.

Communicational Habits and Metaphysical Considerations

I've spoken about hieroglyphs, printed images, photos, and MMS. The ability to display details, create a certain "context", represent something as it is as well as metaphorically, are features all these have in common. Interpreting hieroglyphs, paintings, drawings, and photos as works of art has its own rules. Works of art can be situated in a kind of intellectual milieu which has its special perspective and, as Ivins put it, "its scheme for laying lines".[16] Hieroglyphs emerged from a culture in order to maintain it. To a contemporary expert it was obvious how to interpret a pictorial statement formulated with hieroglyphs. Images and diagrams applied in science are understandable in the context of the given scientific background. MMS is different: it is a new dimension of communication in a culture in which ideas have traditionally been formulated primarily with words; the rediscovery of images is an ongoing process.

In this short and rough comparison of the metaphysics of pictorial and alphabetical writings concentrating on intellectual attitude, I did not touch upon the social or political aspects of writing, details of the development of duplicating images, the conceptual changes or the changing style of verbal expression, etc. I hope that despite these insufficiencies, it has become clear that metaphysical presuppositions and everyday practice are not as far from each other as it used to be supposed, and, more importantly from the point of view of the main issue of the Budapest conference, that although we have to learn to avail ourselves of the possibilities opened by multi-medial communication, we have already taken the first steps.

---

[16] Ivins, *op. cit.*, p. 164.

Seinsvergessenheit, Bildsvergessenheit: **Images in Changing Media***

The importance of images in scientific as well as everyday life is hardly questionable.[1] However, the dissemination of images has never occurred with the same ease as recently. Since the exact duplication of images became possible rather late in historical terms, images as a concise and effective tools were absent from communication. As images became duplicable, their use gradually increased: e.g. as illustrations and diagrams in different kinds of publications[2], as a sign-system used in public spaces[3], and nowadays as an important part of the interface between technology and its users.[4] Recently, a new invention in communications technology, multimedia messages (MMS), reveals an incapability of using pictorial messages in mobile communication. That is, despite the fact that the technological device is ready at hand, in the early days of MMS one could notice a certain kind of inability to use images confidently in its unambiguity and with the same ease as in the case of written messages. Paraphrasing Heidegger, the term *Bildsvergessenheit* (the oblivion of images) refers to this inability. However, there is a close and deep connection between *Seinsvergessenheit* (the oblivion of Being) and *Bildsvergessenheit.*

---

* This is a revised version of a talk given in September 2004, in Krems, Austria, at the Zentrum für Bildwissenschaften conference of *Bildgedächntis – Bildvergessen. Survival of the Images.*

[1] As regards scientific thinking I have in mind the numerous specific examples of scientific discoveries; and then recall the different kinds of instructions illustrated by pictures – to mention just some obvious cases.

[2] About the importance of illustrations and diagrams in scientific development see Elisabeth L. Eisenstein, *The Printing Press as an Agent of Change: Communications and Cultural Transformation in Early-Modern Europe* I-II. Cambridge: Cambridge University Press, 1979. Compare also: "The only knowledges in which the Greeks made great advances were geometry and astronomy, for the first of which words amply suffice, and for the second of which every clear night provides the necessary invariant image to all the world." (William M. Ivins, Jr., *Prints and Visual Communication*, Cambridge, Massachusetts: Harvard University Press 1953, p. 16.)

[3] I think of Otto Neurath's ambition to create an International System of Typografic Picture Education, "an independent and interconnected system of images, to be used together with *word languages*, yet having a visual logic of its own… The icons elaborated within the framework of the isotype program have served as models for those international picture signs we today daily encounter at airports and railway stations". (Quoted by J. C. Nyíri, "The Picture Theory of Reason", in B. Brogaard and B. Smith (eds.), *Rationality and Irrationality*, Vienna, 2001, pp. 250 f.)

[4] Ilpo Koskinen, Esko Kurvinen, Turo-Kimmo Lehtonen, *Mobile Image*. Edita Prim Inc. 2002, p. 13.

To clarify the connection between Heideggerian criticism and the oblivion of images, first I would like to sketch the impact of the media on cognitive and intellectual habits, and then outline Heidegger's main achievements comparing them with ancient Egyptian thought which was formulated within the framework of pictorial writing. By way of conclusion, I'll summarize the main differences between the pictorial and the alphabetical mind, and try to show that a shift from a linear and dualistic way of thinking to a holistic and multimodal one has already begun.

Changing Media – Changing Intellectual Framework

Investigations by Harold Innis, Walter J. Ong, Eric A. Havelock, and William M. Ivins, Jr. imply that the medium in which ideas are formulated is of great importance with regard to cognitive capacity and intellectual disposition.

The Ongian terms of primary orality, literacy, and secondary orality denote the three main epochs which are accessible to us either due to the tokens of remembrance or live experience. In the age of primary orality, a special language was needed to be able to preserve knowledge without recording it. That is, with the help of rhythm, formulas, redundancies, and rhyme it was possible to preserve information through live intercourse. Whereas this intercourse was participatory and agonistic – integration of new experiences was considerably limited. After the emergence of alphabetical writing the abovementioned technical limits fade away, but at the same time, new ones emerged. The orienting situation, gestures, and facial expressions had to be expressed by words; and a need to be able to express general and eternal truths appeared. Due to this demand, the general agent and abstract concepts emerged. As Ivins noted, the literate mind thought in generalities because there was no proper means to express particularities.[5] As there was no proper technology for duplicating images, they lost their informative force.[6] Of course, images have some importance and usefulness in the world of manuscripts as well, as later in the age of the printing press. Earlier images were used to help the glance find the right

[5] Ivins, *Prints and Visual Communication*, p. 91. (In the present volume pp. 49 f.)

[6] Cf. Ivins's comment: "Plato's Ideas and Aristotle's forms, essences, and definitions, are specimens of this transference of reality from the object to the exactly repeatable and therefore seemingly permanent verbal formula. An essence, in fact, is not part of the object but part of the definition. Also, I believe, the well-known notions of substance and attributeable qualities can be derived from this operational dependence upon exactly repeatable verbal descriptions and definitions – for the very linear order in which words have to be used results in a syntactical time order analysis of qualities that actually are simultaneous and so intermingled and interrelated that no quality can be removed from one of the bundles of qualities we call objects without changing both it and all the other qualities." (Ivins, *Prints and Visual Communication*, p. 63.)

locus, then to illustrate verbally expressed experiences. Since the duplication of images was troublesome and the outcome was unsure in terms of exactness, images couldn't be used as the standard pattern/definition of a given thing. After the invention of printing technology, images had to be interpreted in some way, i.e. they have to be redesigned in accordance with the given technology. Moreover, each master has his own syntax of lines and perspective.[7] Accordingly, after the invention of the printing press the informative value of a picture still remained in shadow.

> Another way we can see that the informational capacity of the print was not realized for a long time, is by noting how many woodcuts were daubed up with colour carelessly applied in such a way as to cover and obscure their lines. So far as their buyers were concerned prints were just pictures and not a special kind of pictorial statement that could be exactly repeated.[8]

Until the emergence of photography, the richness and importance of the details of an image remained unnoticed. In the age of secondary orality, technological inventions can create situations which perform similarly to some institutions of primary orality such as "participatory mystique, its fostering of communal sense, its concentration on the present moment, and even in its formulas". However, this orality relies heavily on the inventions of literacy.[9] During this time, the technology of photography also emerged, but as Ivins points out, despite the slow development of the technology of photography,

> we find ourselves in the peculiar dilemma of having a technical knowledge and capacity that are far in advance of many of our settled, accepted modes of thought and valuation, which have remained just as they were before even initial steps were taken towards photography and are based on notions that in many respects are incompatible with its modern developments. I know that this is true in what is called art, and I have a suspicion that it is true in much of academic philosophy also.[10]

That is, the mode of thinking still remained in accordance with the requirements of the literate mind.

These apparently technical changes have had a dramatic impact on thinking. In the course of the history of philosophy, significant shifts occurred in accordance with changes in communications technology: the main issues of philosophy, the ontological and epistemological constructions, the concept of language, etc. can be considered as answers to the challenge of changing media.

---

[7] See Ivins, *Prints and Visual Communication*, p. 61.

[8] Ivins, *Prints and Visual Communication*, pp. 28 f.

[9] Ong, *Orality and Literacy: The Technologizing of the Word*, London: Methuen, 1982, p. 136.

[10] Ivins, *Prints and Visual Communication*, p. 134.

Ivins regards Western thinking as the outcome of a special technological inability.

> I have a notion that much of the philosophical theory of the past can eventually be traced back to the fact, that whereas it was possible after a fashion to describe or define objects by the use of arbitrary and exactly repeatable word symbols addressed, mediately or immediately, to the ear, it was not possible to describe or define them by exactly repeatable images addressed to the eye.[11]

The Heideggerian Criticism

Western philosophy gradually closes itself up in the world of texts. Nonetheless, by the 20th century, communication was equipped with the technologies of photography, film, and various means of telecommunications. Casting a glance at 20th-century metaphysics, we may notice that as new equipment in communications technology became capable of directly mediating numerous moments of everyday life, metaphysics discovered everyday practice and tried to integrate it into its framework.

Against the background of the sterile analyzing tradition of Western metaphysics,[12] Heidegger targeted re-uniting human beings and their environment, focusing on their commitments and activity (activity in the sense of *care* and *work*). I won't delve deeply into the details of Heideggerian philosophy, but I will attempt to sum up his criticism of Western metaphysics and to explicate a key notion of his philosophy, *alétheia* (disclosedness, unconcealment), as it is closely related to disclosure, the distinguished role of thinkers and poets.

Heidegger emphasized our *thrownness* into the world as opposed to the traditional dualism of subject and object. The world, which was earlier considered as an object to be known and separate from the human intellect, emerged as a *referential totality* helping our orientation. Heidegger speaks about a certain forgetfulness which led us astray in our comprehension of reality, the human being and Being. Consequently, key terms were erroneously translated, concealing their original meaning.

---

[11] *Ibid.*, p. 62.

[12] See the description of theoretic culture by Donald: "[T]heoretic culture broke with the metaphoric style of meaning in oral-mythic culture. Where narrative and myth attribute *significances,* theory is not concerned with significance in the same sense at all. Rather than modeling events by infusing them with meaning and linking them by analogy, theory dissects, analyzes, states laws and formulas, establishes principles and taxonomies, and determines procedures for the verification and analysis of information. It depends for its advanced development on specialized memory devices, languages, and grammars." (Merlin Donald, *Origins of the Modern Mind. Three Stages in the Evolution of Culture and Cognition,* Cambridge: Harvard Univ. Press 1993, pp. 274 f.)

This process has led to the dominance of a special one-sided thinking which is determined by the analyzing attitude necessary for theoretical judgments. *Logos*, the ancient term which was earlier understood primarily as a certain revealing force, became logic. Heidegger does not deny that language can be an object of theoretical investigation as the different phenomena of nature are for science, i.e. language can be a present-at-hand (*vorhandenes*) thing. Language can be seen as ready-at-hand (*zuhandenes*) in connection with the practice whereby we articulate our comprehension "Being-with others". In this case, language is an instrument of communication. But Heidegger stresses that language is also a constitutive element of being-in-the-world, since it creates an important moment of revealing.

In the case of *alétheia*, Heidegger stresses the importance of the play of concealment and un-concealment. Heidegger demonstrates the difference between the ancient Greek and recent meanings of the same term. He refers to the Epicurean proverb "λαθε βιωσαζ".

> We translate in 'correct' German: 'Live unnoticed'. But the Greeks say: 'Be concealed in the way you conduct your life'. Here concealment determines the character of the presence of man among men. The 'concealed' and the 'unconcealed' are characters of the very being itself and not characteristics of the noticing or apprehending. Nevertheless, perceiving and saying have indeed for Greeks, too, the basic feature of 'truth' or 'untruth'.[13]

The slight modification in emphasis indicates an unreflected shift: the less accessible the very being itself and its activity, the more dominant the perspective of apprehending. That is, since for the Greeks there was no difference between "perceiving and saying", the difference between beings and words was not significant. "Unconcealment is, so to speak, the element in which Being and thinking and their belonging together exist. *Alétheia* is named at the beginning of philosophy, but afterward it is not explicitly thought as such by philosophy."[14]

This omission leads to "[t]he incapacity of our way of seeing things that is here coming to light, the inability of our thought to experience the unifying unity in the essence of language".[15] Heidegger clearly describes how language becomes a representational means and at the same time how it becomes less and less accessible. He refers to Aristotle who embraces language as speech.

---

[13] Martin Heidegger, *Parmenides* (1982), transl. by André Schuwer and Richard Rojewicz, Bloomington: Indiana University Press, 1998, p. 24.

[14] Martin Heidegger, "The End of Philosophy and the Task of Thinking" (1964), in: D. Farrell Krell (ed.), *Martin Heidegger. Basic Writings*, London: Routledge, 2007, pp. 445 f.

[15] Martin Heidegger, "The Way to Language" (1959), in: *Martin Heidegger. Basic Writings*, p. 407.

Letters show sounds; sounds show affections in the soul; affections show the matters impinge on us... [S]howing brings something to appear, lets what appears be apprehended, and enables what is apprehended to be thoroughly discussed... However, the kinship of the showing with what it shows never unfolds purely in terms of the kinship itself and its provenance. In subsequent periods the kinship is transformed into the conventional relationship between sign and its signified. Greek civilization at its acme experiences the sign on the basis of showing, the sign having been coined by showing for showing. From the Hellenistic (and Stoic) period onward, as the convention becomes sheer stipulation, the sign comes to be an instrument for designating; by means of such designation, representation is coordinated and directed from one object to another. Designation is no longer a showing in the sense that it lets something appear.[16]

A bit later, Heidegger continues the above thought. He writes: "*What unfolds essentially in language is saying as pointing*. Its showing does not culminate in a system of signs. Rather, all signs arise from a showing in whose realm and for whose purposes they can be signs."[17]

As regards the distinguished role of poets and thinkers, revealing again has particular role. Both engagements imply a special relation to language. Philosophy, whose "ultimate business ... is to preserve the *force of the most elemental words* in which Dasein expresses itself",[18] attempts to show the work of *alétheia* in its entirety. On the other hand, as Richardson put it, "the poet's task would be to give a name to Being in its positivity (as the Holy)".[19] This positivity stresses "unconcealment" in a peculiar way.

The poet calls, in the sights of the sky, that which in its very self-disclosure causes the appearance of that which conceals itself, and indeed *as* that which conceals itself. In the familiar appearances, the poet calls the alien as that to which the invisible imparts itself in order to remain what it is – unknown.[20]

---

[16] *Ibid.*, p. 401.

[17] *Ibid.*, p. 410. Although Heidegger emphasizes here that "in view of the well-joined structure of the saying, we dare not attribute showing either exclusively or definitely to human doing" (*ibid.*), in another work he actually stresses the importance of "human doing". "Man himself acts [handelt] through the hand [Hand]; for the hand is, together with the word, the essential distinction of man." (Heidegger, *Parmenides*, p. 80.) Some pages later when he considers the effect of the typewriter (and also the concomitant movements of the hand) and printing press he concludes: "when writing was withdrawn from the origin of its essence, i.e., from the hand, and was transferred to the machine, a transformation occurred in the relation of Being to man." (*Ibid.*, p. 85.)

[18] Martin Heidegger, *Being and Time* (1927), transl. by John Macquarrie and Edward Robinson, New York: HarperCollins, p. 262.

[19] William J. Richardson, S.J., *Heidegger. Through Phenomenology to Thought*, The Hague: Martinus Nijhoff, 1967, p. 545.

[20] Martin Heidegger, "...Poetically Man Dwells...", in: *M. Heidegger, Poetry, Language, Thought*, transl. by Albert Hofstadter, New York: Perennial Classics, 2001, p. 223.

Modifications in the history of philosophy (just like in the case of technology) reveal the important turning points in philosophical thinking.[21] These modifications show the basic attitude toward beings and Being, and basic convictions which make up the dominant world view. In this sense, authentic philosophical investigation is capable of displaying both concealment as the emerging problems become covered by provisional solutions; and at the same time unconcealment, after all problems come to light. As Heidegger believes, "[e]mergence into the unconcealed and submergence into concealment dwell primordially everywhere".[22] Poets, on the other hand, show things and constellations as they are, and this showing reveals the mysterious, the inaccessible as it is, without the distorting effort which tries to reach and grasp it.

This special capability is due to the fact that poetry "speaks in images".

> The poet makes poetry only when he takes the measure, by saying the sights of heaven in such a way that he submits to its appearances as to the alien element to which the unknown god has 'yielded'. Our current name for sight and appearance of something is 'image'. The nature of the image is to let something be seen. By contrast, copies and imitations are already mere variations on the genuine image which, as a sight or a spectacle, lets the invisible be seen and so imagines the invisible in something alien to it. Because poetry takes that mysterious measure, to wit, in the face of the sky, therefore it speaks in 'images'. This is why poetic images are imaginings in a distinctive sense: not mere fancies and illusions but imaginings that are visible inclusions of the alien in the sight of the familiar.[23]

And, as Heidegger suggests, image has a particular capability as opposed to language. Recalling the modification of sign, the distancing, alienating character of language emerges. Due to the transformation of sign into a means of designation, words are not capable of calling to presence the thing whose sign/name they are, rather words just signify, refer to the thing itself; "designation, representation is coordinated and directed from one object to another".[24] As opposed to this one-to-one correspondence, images are capable of showing particularities and similarities at the same time. Heidegger distinguished three different senses

---

[21] "Descartes established man as the ontological center: henceforth, the reality of the real was what *man* asserted to be. Recall that for the Greeks, the reality of the real was constant presence. For medieval man, too, presence was determinative for reality: God's absolute presence maintained the existence of all His creatures. With Descartes, the reality of the entire cosmos was also revealed as presence, but as something present *for* the self-certain subject (man)... Because his own self-consciousness is the standard by which he judges the truth of everything else, what is true is what human rationality determines to be true." (Michael E. Zimmerman, "Beyond 'Humanism': Heidegger's Understanding of Technology", in: Thomas Sheehan (ed.), *Heidegger. The Man and the Thinker*, Chicago: Precedent Publ. Inc. 1981.)

[22] Heidegger, *Parmenides*, p. 67.

[23] Heidegger, *Poetry, Language, Thought*, p. 223.

[24] *Ibid.*

of the term 'image' by Kant, and I believe these senses are also valid for the everyday usage of the term: "immediate look of a being, as the at-hand, likeness-taking look of a being, and as the look of something in general."[25] As he writes a bit later: "In the one which applies to many, they [images in the broadest sense] show how something appears 'in general'".[26]

Pictorial Writing and Iconic Philosophy

Ancient Egyptian pictorial writing and the mythological-philosophical doctrine attached to it are in harmony with Heidegger's endeavour: Egyptian thinking orders everyday experiences into a holistic framework where each phenomenon, its representation, and the act of representation are the same – just as in the case of the revealing power of poetry described by Heidegger.

Pictorial writing, as the ancient Egyptian way of *monumental writing*[27] can be considered, is a special combination of hieroglyphs and images. Hieroglyphs as signs can represent sounds, i.e. they can function as phonograms, can represent something similar (can function as ideograms), and can be taxograms /determinatives which are used to determine different phenomena on the basis of the kind of abstractness. Some hieroglyphs bear all three of these functions and the given combination of signs makes the actually valid function clear, i.e. indicates how it is to be interpreted.

The Egyptian symbol system has another special characteristic: the border between images and hieroglyphs is not really sharp, and moreover, hieroglyphic writing is at the same time a fine art. According to Jan Assmann's theory, fine arts and hieroglyphs can be identical as a means for making manifest God's words with special regards to the maintenance and preservation of culture and identity.[28] Similarly, image and hieroglyph are not easily distinguishable from a technical-visual perspective. The reason for this is that pictorial representation is formal, symbolic, and geometrical, accordingly hieroglyph-like, and hieroglyphs

---

[25] Martin Heidegger, *Kant and the Problem of Metaphysics* (1929), transl. by Richard Taft, Bloomington: Indiana University Press, 1997, p. 65.

[26] *Ibid.*, p. 67.

[27] I refer here to the terminology of Jan Assmann. He distinguishes hieroglyphic writing from cursive, i.e. hieratic script as the latter was designed for everyday purposes opposed to the former which was used to maintain "cultural memory". Jan Assmann, *Das kulturelle Gedächtnis: Schrift, Erinnerung und politische Identität in frühen Hochkulturen*, München: C. H. Beck, 1992, pp.170 f.

[28] *Ibid.*, p. 170.

are quite often used as images, ie. developed hieroglyphs replace images of the given phenomena of a relief or fresco.[29]

This flexible recording system is suited to formulating and maintaining a special cult and culture, which seems to be quite strange from the perspective of Western rationality. In my outline of some of the main characteristics of Egyptian philosophy, I rely considerably on a book by a Hungarian scholar, Attila M. Farkas.[30]

One distinctive speciality of Egyptian thinking is its "complementary logic". As opposed to the Western standard, contrary features or states of affairs are not excluding but rather complementary, completing elements of the whole. The cyclic emergence of the Sun clearly enables things to become visible, humans become capable of seeing them and understanding the state of affairs and being active. After sunset all these capabilities are eliminated, suspended while it is time for sleeping and dreaming, the netherworld, and the original chaos before creation. Sunrise and creation, just like any kind of human capability, are connected.

This interrelatedness of different kinds of things (like natural phenomena, creation, and capabilities) is easily expressible through pictorial signs which preserve feedback to their roots. It enables flexible interpretations in accordance with the given circumstances, thus it is possible to interpret pictorial writing either directly or metaphorically. Being able either to represent or understand something means either the creation of or an immersion in a given situation. That's why it is necessary to show each important feature of the represented scene, and not only those which are visible from one perspective.[31]

Reality, creation (both in religious and everyday senses), and the recording and understanding of recorded statements belong together.[32] Words and forms through which it is possible to express understanding are inseparable from each other. Because of the close interrelatedness of the elements of reality and their creation and mental processing, there is no separation between name, sign, and the meaning.[33]

---

[29] Attila M. Farkas, *Filozófia előtti filozófia. Szimbolikus gondolkodás az ókori Egyiptomban* [Philosophy Before Philosophy. Symbolic Thinking in Ancient Egypt], Budapest: Typotex 2003, pp. 163 f.

[30] For a more detailed explication see the chapter above, "The Iconic Turn in Metaphysics".

[31] Compare this with the invention of 15th-century painting, the foreshortening which tried to represent things as they are visible from a given angle.

[32] See the relevant details of Memphite Theology in the previous chapter: pp. 97 f.

[33] Farkas, *Filozófia előtti filozófia*, p. 266.

The Adaptability of the Literate Mind

In Western thinking, this unity has disintegrated. This change is obvious both in light of the difference between pictorial and alphabetical writing, and in Heidegger's criticism. According to the former, pictorial writing bears a creative force. Alphabetical writing merely records the acoustic phenomena of speech via arbitrary signs; it can't have the creative potential that language has been supposed to have because, as Heidegger pointed out, "sign comes to be an instrument of designating... Designation is no longer a showing in the sense that it lets something appear."[34] Accordingly, sign and meaning, represented things and relations, can't be the same. Reality and its cognitive-cultural processing are separated from each other. As Heidegger puts it:

> The Greeks experience and express concealment in many ways, not only within the sphere of the everyday handling and considering of things, but also from the ultimate perspective of beings as a whole. Death, night, day, the earth, the subterranean are pervaded by disclosure and concealment and remain mired in this essence. Emergence into the unconcealed and submergence into concealment dwell primordially everywhere.
>
> This dwelling of disclosure and concealment, in advance and everywhere and always and for every being and in all Being is expressed in the word of the Greeks. It is what is said primordially – the legendary [*Sage*, Saying].[35]

The similarities between the holistic view of the world of ancient Egyptian thought and ancient Greek philosophy as described by Heidegger are conspicuous. The creative potential of language is due to its revelatory force, which means an intimate relationship between words and things, as Heidegger described. Pictorial signs are capable of maintaining this intimate relationship because of the visible roots of the pictorial signs, and because of the capability for flexible usage. I suggest there is considerable similarity between the capabilities of Heidegger's poets and of Egyptian monumental writing.

As regards Western metaphysics, Heidegger calls attention to the one-sided commitment to an analyzing attitude and, as a consequence, to language which is ruled by logic. He attempts to demonstrate an alternative perspective from which it is possible to restore the earlier disrupted unity of human beings and their environment, and overcome the dominance of language which has degenerated into a designating mechanism. These two attempts are obviously the same, since restoring the unity of reality and the cognitive-intellectual processing of it, is only possible if generalities and particularities are manageable within the same frame of reference. Within the framework of the traditional dualistic approach, the particularities of everyday life and general princi-

[34] See the quotation on p.108 above.

[35] Heidegger, *Parmenides*, p. 67.

ples of sciences and philosophy were not manageable coherently.[36] It becomes apparent in Heidegger's deliberations on language that the main means of clear-cut alphabetical reason, i.e. written language, is not capable of mediating everyday experiences and an abstract comprehension of reality as a whole. Conversely, pictorial writing in ancient Egypt proved to be an appropriate means for mediating both generalities and particularities.

During the 20th century, the informative force of images was gradually discovered; nowadays images are effective and useful mediators of different functions and experiences, as well as helpful in the documentation of experiments and in illustrations of problem-solving. The theoretical re-discovery of images has already started, not only from the perspective of art history but also from the point of view of philosophy. Against the background of the literate mind, these achievements of Western rationality are indeed promising.

[36] I have in mind the dominant attempt to describe reality through general principles, and these principles mostly have nothing to do with everyday activities. As an exceptional example, see the topic of friendship in chapter "Social Makeup and Solitary Reason".

The Mobile Image: Experience on the Move*

As visuality is an important part of everyday practice, pictorial representation is also an essential element in everyday communication. McLuhan's famous aphorism that "the medium is the message" has long been discussed within the paradigm of orality vs. literacy. Nowadays, when within the practice of everyday telecommunications in the narrow sense, images are gaining ground, it is worth investigating the communicational potential of images from the perspective of this paradigm.

On the one hand, we consider the communicational potential of images to be their natural capability; on the other hand, however, we are not quite ready to use them confidently. There are deeper reasons in cultural history for this duality. The long centuries of literacy gave rise to peculiar cognitive techniques that involve the conceptual processing of experience in a way which does not completely overlap with the result of visual information processing. The difference between these cognitive techniques and visual information processing sheds light on the very feature of images which makes their utilization in everyday telecommunications and particularly in mobile communication a prospective agent of important changes.

In his book *Prints and Visual Communication*, William Ivins puts forward a convincing summary of the technical and technological reasons for, and the cognitive consequences of, the century-long oblivion of images. His historical overview reveals that, as it took a long time to reach a reliable method of reproducing pictures, the potential of images to mediate information was basically neglected in thinking, adapting to the constraints of verbal expression. Moreover, this remained the dominant attitude and condition even in the decades following the rise of the photograph.[1] However, time has not passed idly since the emergence of photography. In fact, efforts to reconquer the realm of images have been made by way of philosophical reflection as well as through the possibilities offered by technology in everyday practice. Here I mean not only the moving pictures mediated by films and particularly television, but also the development and diffusion of various systems of iconic signs (such as ISOTYPE

* First published in Kristóf Nyíri (ed.), *Mobile Studies. Paradigms and Perspectives*, Vienna: Passagen Verlag, 2007, pp. 25-33.

[1] "Thus we find ourselves in the peculiar dilemma of having a technical knowledge and capacity that are far in advance of many of our settled, accepted modes of thought and valuation, which have remained just as they were before even the initial steps were taken towards photography and are based on notions that in many respects are incompatible with its modern developments." (William M. Ivins, Jr., *Prints and Visual Communication*, Cambridge, MA: Harvard University Press, 1953, p. 134.)

or a multitude of emoticons, which are now commonly used, especially in mobile communication).

Similarly, the study of the role played by images and pictorial messages in mobile communication can no longer be considered to be an intact terrain. For example, several notable findings, including some within the narrower field of research on MMS, have been published within the framework of the cross-national interdisciplinary research programme *Communications in the 21st Century*, launched in 2001. At an early stage, in 2002, Kristóf Nyíri had already pointed to the essential role of images in thinking and the possibilities deriving from the fact that pictures have become easy to communicate.[2] Since then, a number of studies have been prepared on the role of visualization and multimodal tools in education,[3] the informative and integrative potential of images, and the distinct characteristics of pictorial representation.

When the use of MMS is studied against the background of photography and its cultural-historical traditions, it turns out that the institution of MMS incorporates photography into ongoing discourses. Findings from earlier research[4] clearly show that the role and function of photographs in our lives has undergone significant transformation. Traditional photos served as memory banks, perpetuating important moments, locations, people, and groups of people. Cameras built into mobile phones have created a new possibility – the possibility of maintaining perpetual visual contact. Since pictorial messaging has become mobile, there are no practical obstacles to the development of "an interactive social context" for photography.[5]

More detailed study of the MMS communication genre brought forth certain typologies. For example, researchers distinguished some general genres of use, separating the documentation of particular situations from the visualization of specific details, or snapshots (which serve to develop camaraderie among colleagues), from postcards and greetings for sharing personal experiences, and "chain-messages" that can generally characterize the communicative practices of smaller or larger communities. These genres have been

---

[2] Kristóf Nyíri, "Pictorial Meaning and Mobile Communication", in Kristóf Nyíri (ed.), *Mobile Communication: Essays on Cognition and Community*, Vienna Passagen Verlag, 2002, pp. 157-184.

[3] See Lynn V. Marentette: "Thinking, Learning and Communicating through Multimedia", in Kristóf Nyíri (ed.), *A Sense of Place. The Global and the Local in Mobile Communication*, Vienna: Passagen Verlag, 2005, pp. 319–328. Multimodal tools can satisfy requirements which may be set by and for adults more efficiently than paper-based education; more specifically, visualization can greatly facilitate understanding in general terms.

[4] Ilpo Koskinen, Esko Kurvinen, Turo-Kimmo Lehtonen, *Mobile Image*, Helsinki: Edita Publishing, 2002.

[5] Ilpo Koskinen, "Seeing with Mobile Images: Towards Perpetual Visual Contact", in *A Sense of Place*, pp. 339–347.

embedded into the existing practice of communication as useful supplements to traditional modes.[6]

Several studies emphasize the integrative power of MMS, which partly derives from the fact that it allows us to share our visual experiences continuously and thus create a shared code and practice of interpreting visual experiences.[7] Some authors, including Barbara Scifo, consider the exchange of visual information to be no less than an act of trust,[8] which can help us, among other things, describe and even testify to situations, allowing those situations and experiences to be shared.

A variety of surveys revealed that young people consider sending a pictorial message to be a very convenient, timesaving, and effective means of exchange;[9] moreover, mobile phones capable of storing images served as photo albums for their users.[10] Empirical studies[11] confirm that MMS exchanges can be mostly found within narrow circles of close friends. This is no coincidence, since the interpretation of an image requires shared contextual information. On the other hand, as almost all analyses based on empirical research emphasize, pictorial messages are very effective tools for maintaining co-presence and visual presence. Therefore the mobile phone, which was already considered a person-specific (personally tailored) device, can help maintain personal presence and reinforce the contacts of its user when it utilizes visuality as a supplementary tool.

The potential of images on the move may also bring about changes in relationships with a community and the location specific to that community. Shared images, functioning as shared experiences, may, on the one hand, make us aware of multiple viewpoints, and, on the other hand, can strengthen ties among members of a community.[12]

The theoretical issue of the difference between verbal and pictorial communication also attracted the attention of researchers. An interesting experiment

---

[6] Rich Ling, Tom Julsrud, "Grounded Genres in Multimedia Messaging", in *A Sense of Place*, pp. 329–338.

[7] Cf. Koskinen, "Seeing with Mobile Images"; Barbara Scifo, "The Domestication of Camera-Phone and MMS Communication", in *A Sense of Place*, pp. 363–373.

[8] Scifo, *op. cit.*, p. 368.

[9] Scifo, *op. cit.*, p. 371..

[10] See also Virpi Oksman, "MMS and Its "Early Adopters" in Finland", in *A Sense of Place*, pp. 349–361.

[11] See, for example, Nicola Döring, Christine Dietmar, Alexandra Hein, Katharina Hellwig, "Contents, Forms and Functions of Interpersonal Pictorial Messages in Online and Mobile Communication", in Kristóf Nyíri (ed.), *Mobile Understanding: The Epistemology of Ubiquitous Communication*, Vienna: Passagen Verlag, 2006, pp. 197–207.

[12] Cf. Fumitoshi Kato, "Seeing the 'Seeing' of Others: Environmental Knowing through Camera-Phones", in *Mobile Understanding*, pp. 183–195.

is described in a paper which analyzes findings for the Hungarian "MMS of All Knowledge" project, where the abstracts of lectures given as parts of the University of All Knowledge/ENCOMPASS series were sent as MMS messages.[13] Certainly, figures and diagrams have had a role in the development of science. Drawing on Allan Paivio's *double coding* theory and certain considerations raised by Wittgenstein, the experiment clearly supported the idea that, depending on content, pictorial and verbal representations can effectively complement each other.

In another article analyzing the origins of pictorial and verbal communications from a phylogenetic perspective, Kristóf Nyíri concludes that images constitute a natural environment for thinking.[14] Images provide a more natural medium than language as far as they can be considered natural signs due to their resemblance to the objects and situations represented. During the long centuries of literacy, the lack of a proper technology for duplicating images basically made people forget the capability of pictures to carry information. However, this situation has changed radically. MMS may – in fact, by today it certainly does – turn images into a ubiquitous means of communication.[15] The potential of pictorial communication to strengthen communities is considerable, even if we take account of the phylogenetic shifts in communication. In a later paper where he contemplates the relationship between images and space,[16] Nyíri reaches a similar conclusion. Unlike words, pictures have an inherently spatial organization and they are much more suitable for representing objects and events. On the other hand, a certain spot in space, namely one's home, has a prominent role. Drawing on Heidegger's trichotomy of building–dwelling–thinking *(Bauen Wohnen Denken),* Nyíri points to the possibility of handing down local tradition. Utilizing images in mobile communication allows the local and personal to be communicated.

We can find remarkable divergence in opinions about the combination of image and text. While Döring et al. emphasize that most pictorial messages sent via either e-mail or MMS do not contain text at all, which is, they add, "an indication of our skills in visual communication"[17], some authors contributing to the subsequent volume for the *Communications in the 21st Century* programme

---

[13] Viktor Bedő, "MMS of All Knowledge (MMSAK): Mobiles and Scientific Visualization", in *A Sense of Place,* pp. 383–393.

[14] Kristóf Nyíri, "Pictorial Meaning and Mobile Communication", in Kristóf Nyíri (ed.), *Mobile Communication: Essays on Cognition and Community*, Vienna: Passagen Verlag, 2003, pp. 157–184.

[15] Cf. the article published on November 30, 2006 by the Hungarian portal Index, which reports on the significant spread of MMS use in Hungary, accessible at http://index.hu/tech/mobil/mobsur2egy.

[16] Kristóf Nyíri, "Images of Home", in *A Sense of Place,* pp. 375–381.

[17] Nicola Döring, Christine Dietmar, Alexandra Hein, Katharina Hellwig, *op.cit.,* p. 207.

published a year later in 2005, including Bedő, Koskinen, and Nyíri, emphasize that image and text tend to be complementary. However, this difference should not come as a surprise if we consider, on the one hand, that it took decades to rediscover images, and, on the other hand, that there are conceptual and intellectual differences between the cultures expressed through images and words (as highlighted in two preceding chapters). Again, this shows well our oblivion of the image: the fact that it is obviously an easier task to process images at an elementary level than to visualize certain information, i.e. represent it in a pictorial form. We possess the former as an innate skill, while the latter is essentially culture-specific.

Research on MMS use has clearly shown that MMS can function as a very informative message carrier, provided that an unambiguous context exists or communicating partners have a common background which allows a certain pictorial language to develop. Therefore, an important condition of success in interpreting images is that the frame of reference including the message should be evident or familiar. A closely related factor is that pictorial messages are very useful in creating and binding smaller communities, as well as in developing closer and more intensive contacts. Pictures are rich in detail and are able to mediate interrelations in their entirety, which, in addition to the characteristics of visual perception, enables images to provide an impression similar to primary experience, i.e. they virtually communicate the direct experience.

It is sometimes said that the image, being a prominent means of mediating reality, has an immersive power. However, such a proposition usually does not define the property by which the image assumes such power. In what follows, I will attempt to give an explanation of this peculiar power possessed by the image, and pinpoint the place of mobile telephony among the agents and modes of mediating representation with respect to mediation and access to experience.

We can approach the source of the image's immersive power via certain features of visual perception. According to the representational theory of the visual mind, the processing of visual information can be described – both functionally and, of course, anatomically – on the basis of a dual model. This theory, developed by Pierre Jacob and Marc Jeannerod,[18] makes a distinction between visual and visuomotor representations. The former provides information for higher-level cognitive functions, such as categorization, conceptual thinking, or argumentation, while the latter serves action.[19] None of these representations has conceptual content, although both can serve as the basis for or be subject to conceptual processing. The authors argue that the content of visual perceptual representations is "both more fine-grained and informationally richer

[18] See Pierre Jacob and Marc Jeannerod, *Ways of Seeing: The Scope and Limits of Visual Cognition*, Oxford University Press, 2003, rev. ed. 2004.

[19] *Ibid.*, p. 45.

than the conceptual contents of thoughts."[20] This means that information gained from visual perception can be transformed into thoughts with a considerable loss of information only, since such information is much more manifold and detailed than something we can grasp conceptually and express as a thought. So, for example, to represent the relation between two objects through concepts, we have to omit several bits of information which are essential parts of the information gained through visual perception. The relation between two objects can be grasped only if we know the pictorial/iconic content (e.g., form, size, colour, etc.) of those objects. On the other hand, when the pictorial content of visual perception is transformed into conceptual content, a new cognitive element, namely a kind of reflexive egocentric perspective, is added to the evolving perception. Consequently, transformation into conceptual content also expresses a reflexively treated relation or situation with respect to the given state of affairs.[21]

In addition to the functional difference between visual and visuomotor representations, there is an essential difference between their frames of reference. Visual representation provides information about the context with respect to the relation between different elements of this context and the order of objects, while visuomotor representation visualizes the elements of the context through the eyes of the perceptor, so as to show the conditions of manipulating objects. Switching between these modes of representation is an automatic act, although it can be hindered by certain brain injuries. However, it is also notable that the egocentric perspective which emerges in the course of conceptual processing considerably differs from the egocentric reference of visuomotor representation. The former is reflexive and can be consciously converted, while the latter is spontaneous and alters only with a change of the actual physical situation.

In visual perception, our pragmatic orientation plays an important role, while the perception of numerous minute details provides us with abundant information about the context. The conceptual processing of these percepts results, at the cost of losing certain pieces of information, in a detached description of the situation. Such detachment is a necessary consequence of the peculiar situation in which the conceptual transformation of pictorial information implies mapping to a considerably narrower domain, and this mapping takes place along the lines of a certain perspective or intention. As Michael Tomasello notes, discussing the special features of linguistic representation in the light of human ontogeny and the difference between primate and human cognition, speaking implies a certain intention and, depending on this intention, it represents a certain viewpoint or perspective. Linguistic representation is built on the

[20] *Ibid.*, p. 22.
[21] *Ibid.*, p. 31.

decision made by individuals in the given situation, i.e. the perspectives that best fit their objectives in interpreting phenomena, rather than the simple recording of various sensory and kinetic experiences.[22]

In contrast, representations which process visual perception are either pragmatic, i.e. specific to action in the given situation, or they basically mediate iconic content which can serve as a basis for various modes of conceptual processing due to its information-richness. The new devices of communication may have a role as agents of these representations. Therefore, utilizing various modes of representation in everyday communication implies a peculiar situation in terms of space and time as well as the reality of the here-and-now. Concerns about potential distortion and manipulation are often shown with respect to mediation and mediated nature. However, as we have seen, a significant change, in fact, distortion, has to be considered even for the most elementary steps of representation. Our implicit, procedural knowledge becomes declarative and consequently accessible and tangible through representational redescription.

> Systems of thought emerge from this reflective activity [i.e., representational redescription] because self-observation employs all of the categorization and analytic skills that are employed in perceiving, understanding, and categorizing the outside world – in effect the subject perceives, understands, and categorizes her own cognition facilitated by the fact that it is expressed *externally* in language.[23]

We can say that linguistic representation itself has a role of mediating between certain kinetic, episodic experiences and communicable cognitive content. Obviously, this mode of mediation within the individual is closely related to the social nature of humans – a fact that is also supported by cognitive evolution research. External storage systems which serve to aid memory, such as writing, helped extend communication in time and space; however, the very possibility of such extension brought about the development of several constraints. As we have seen, verbalization itself means a certain restructuring of our perceptions, and written text has to be self-explanatory without any other sensory supplement to facilitate understanding. This is why making voice transmittable, and later, making images and moving pictures telecommunicable significantly facilitated communication in certain situations. If we also consider that, as we can see nowadays, all these capabilities are becoming an organic part of our everyday lives via portable and user-friendly devices, then apparently the world of the here-and-now is separated from the reality of the there-and-then

---

[22] "[T]he intersubjective and perspectival nature of linguistic symbols actually undermines the whole concept a of a perceptual situation by layering on top of it the multitudinous perspectives that are communicatively possible for those of us who share the symbol." Michael Tomasello, *The Cultural Origins of Human Cognition*, Cambridge, MA: Harvard University Press, 1999, p. 132.

[23] *Ibid.*, p. 195, emphasis not in the original.

experienced by others by a very thin boundary only. Therefore, we can easily share our primary experiences with others, and, in fact, in a form which allows others, through visuality, to access these primary experiences directly.

The transformation of technology from analogue to digital is usually linked to the shift from real to virtual. However, this change is not as radical as we might think at first sight. As Ron Burnett writes, "images are virtual because they are distant from the spectator or user but are experienced as if that distance could and, in some instances, must be overcome."[24] Consequently, the reality represented by pictures excludes, through the power of visual perception, the detachment so characteristic of verbality, and thus an attitude prevails which generally characterizes our relationship with the environment. If we consider how this property – i.e. the one that characterizes our basic relationship with the environment – exerts an effect via the use of MMS and the videophone as a part of our everyday lives, then we can see spontaneous adaptation on the one hand, and feel that we can easily and freely share our experiences on the other. In fact, images may significantly ease the burden of multitasking, which is often attributed to mobile telephony. The ceaseless switching between problems and tasks to be solved, which is a constant duty posed by flexible time management, will become simpler if a problem located somewhere else in space becomes virtually tangible because we can see its context through a real-time account (videophone) or as the result of a pictorial account (MMS). Apparently, the mobile, which has become a part of our everyday lives, is a device increasingly suitable for representing reality, and thus it may become one of the main mediators of our experiences.

How can this practice of mobile communication modify our cognitive habits? How will the habitual relationships between our visual experiences and the surrounding world develop? Will the dominant position of here-and-now reality change? The workload generated by the necessity of performing various tasks simultaneously can decrease if the things that are now distant are brought closer – in a sense, unnoticeably – by the immersive power of images. Furthermore, we can gain in-depth knowledge through the orienting and informative power of images which greatly facilitates the understanding of particular situations /problems. Of course, such a practice of sharing experiences reduces the significance of spatiality. However, we should also ask whether it will have a greater role, provided that particular relationships between experience and locality can evolve through the permanent interplay of the virtual and the real. The contrast between real and virtual may decrease further if they are habitually switched in everyday practice. Recall that this contrast is the product of that very era during which words had an overriding role in communicating thoughts.

[24] Ron Burnett, *How Images Think*, MIT Press, 2004, p. 72.

Multimodal Integration
From a Philosophical Point of View*

In this paper, I argue for the integrative power of mobile devices both from a cognitive and social point of view. (The ease of building new elements into the framework of earlier acquired knowledge is crucially important when considering any kind of learning activity.) As part of my reasoning, I rely considerably on the lessons of cognitive evolutionary studies, and on an overview of the development of communications technology as well as that of the educational system. Changes recently experienced suggest that the slowly modifying habits in everyday activity could cause institutional changes as well, and mobile technology plays an important role in this process. Mobile technology integrates almost all the inventions of the last century into the realm of communications technology, i.e. it is capable of mediating nearly all kinds of experiences at any time and with no regard to its location. Accordingly, it could be a dominant mediator of social intercourse. Since education is tailored according to the limits and demands of literacy, while at the same time it tries to tame new technology, (i.e., integrate it into the framework of traditional patterns), change may not be as fast as expected. Mobile learning could yield a vision of an educational system which is deeply immersed in everyday practice and ready to answer emerging demands.

M-Learning Revisited

Recently, mobile learning has become a blooming research field. It has divergent focuses, ranging from technical to pedagogical details. There are important questions to be answered. How does mobile learning fit into formal education? What is most suited to learning that uses mobile technology, and how can we best adapt it to mobile devices? What should be emphasized: content on the move or the learner on the move? There are many experiments providing evidence that portable devices are useful accessories to certain functions (navigation, interactive audio sets in museums, etc.). This evidence proves the usefulness of mobile devices, although it emphasizes the adaptation of new technology in a traditional environment. An important question arises regarding how to design inventions in communications technology: should they be in accordance with a traditional institutional framework or according to some other vision?

---

* First published in Immaculada Arnedillo-Sanchez (ed.), *Proceedings of the IADIS International Conference. Mobile Learning 2007*, pp. 119-126.

Mostly, innovations in technology focus on adaptability and efficiency in the given circumstances. The development of technology is about the invention that makes certain functions easier and more effective. The dilemma I mention above comes to life because of slight changes in most fields of everyday activity which suggest that an overall change is imminent.

Traditional differentiations of work and spare-time, public and private sphere, learning and post-learning, later on, working, of the workplace and home, theory and practice, and so forth seem to disappear from everyday routine. Teleworking, lifelong learning, and the continuous rescheduling of tasks are part of everyday activities. Epistemology, cognitive science, and the philosophy of science emphasize the importance of the interrelatedness of theory and practice and the mutual work of conceptual and non-conceptual elements. That is, the conceptual processing of these modifications is under process. These changes are not by chance and indicate an overall change in the social, institutional, and even cognitive framework. I recommend investigating mobile learning within the wider horizon of everyday practice as opposed to maintaining a special focus on the learning process. Within this framework, the difficulties of how to decide what features of mobile learning to emphasize in design vanish, and, on the other hand, it frees the way for changes.

Let me shed light on the theoretical background of this suggestion. Changes in everyday practice are closely related to communication, more precisely, to communications technology. Communications technology determines the boundaries of how thoughts and ideas can be expressed, and also what kind of ideas can come to mind. I'd like to recapitulate how this can have an effect on the cognitive patterns of the individual and on the institutional framework of a community/society.

Merlin Donald's theory of cognitive evolution makes the connection between communications technology and cognitive setting clear, as well as the connection between the two pillars of his theory, i.e. social intelligence and the evolution of the representational system. Regarding social intelligence, Donald quotes Robin Dunbars' *social brain* hypothesis,[1] which suggests a direct relation between group size and the size of the neocortex. That is, cognitive capacity has to be increased in accordance with the growth of a primate group, in order for individuals to be able to keep relations in mind and to be able handle these relations. As Donald concludes "it is clear that language was the final step, and

---

[1] This hypothesis suggests "co-variation between neocortex volume of primates and various aspects of primate social behaviour, including social group size, grooming clique size, the use of subtle social strategies such as alliances and tactical deception and the levels of social play. There is, for example, a linear relationship between relative neocortex volume and social group size in primates". (R. I. M. Dunbar, "Are There Cognitive Constraints on an E-World?", in Kristóf Nyíri (ed.), *Mobile Communication. Essays on Cognition and Community*, Passagen Verlag, Vienna 2003, p. 58.)

that presymbolic forms of social intelligence must have been its foundation".[2] Donald distinguishes three main transitions in human cognitive evolution: a shift from episodic to mimetic, then to mythical, and finally to theoretical culture. Each of these changes means the emergence of a new kind of representation.

> Human memory had, from its inception, expanded the range of primate memory. The earliest form of hominid culture, mimetic culture, depended on an expansion in the self-representational systems of the brain and created the initial base for semantic memory storage, which consisted initially of representational action scenarios reflected in mime, gesture, craft, and skill. With the evolution of speech and narrative ability, there were even greater increases in the load on biological memory, adding not only the storage networks for phonological rules and the lexicon in its entirety but also a very large store of narrative conceptual knowledge.[3]

The transition from mythical to theoretical culture presupposes the existence of an effective external storage system.

> [T]he first two evolutionary transitions would have greatly increased the load on biological memory. However, the final step in this tremendous cognitive expansion might have reduced the load on some aspects of biological memory, by gradually shifting many storage tasks onto the newly developed E[xternal]S[ymbolic]S[torage system]. At the very least, the existence of the ESS must have forced a great change in priorities and memory organization.[4]

The transition from mythical to theoretical culture is also a well documented body of scholarship. The so-called Toronto Circle provided plenty of evidence regarding this transition, according to their terminology, from the culture of primarily orality to the age of literacy. Not only has the role of "old wise man" changed, or more precisely been eliminated, but memory technique and legislation, the manner of weaving ideas together, have radically altered too. Since alphabetical writing was very effective in preserving, accumulating, and reproducing ideas, it was able to become the main forum for objectifying ideas. Though writing is a clear and unambiguous way to record, there was a price to pay for its effectiveness: all elements of the idea or narrative had to be expressed in words. There is no supporting environment or unity of references that can help understanding. As a new kind of storage system with built-in bounds, alphabetical writing created new habits in the thought process compared to orality. General subjects, abstract concepts, and linearly structured arguments emerged. The cohesive power of live intercourse and its multimodal conditions

---

[2] Merlin Donald, *Origins of the Modern Mind. Three Stages in the Evolution of Culture and Cognition* (1991), Harvard Univ. Press, Cambridge, 1993, p. 137.

[3] *Ibid.*, p. 319.

[4] *Ibid.*, p. 320.

remain an important factor in everyday life, but the conceptual and institutional framework have gradually been guided by the requirements of mute texts.

Secondary orality set in with the telephone, radio, television, various kinds of audio tape, electronic technologies, and the like, which to some extent enabled the renewal of old institutions and inclinations, as well as new demands. "This new orality", as Ong put it, "has striking resemblances to the old... But it is essentially a more deliberate and self-conscious orality based permanently on the use of writing and print."[5] Recently, emerging new communications technology has yielded instant access to different kinds of external storage systems. Since this storage is multimodal and accessible at any time from nearly anywhere, the arrangement of knowledge and the institutions of knowledge-acquisition tailored by the requirements of literacy should be changed.

The Integrative Potential of Mobile Technology

First of all, I'd like to sum up what mobile technology provides for us. A well-researched and well-documented body of scholarship indicates that mobile devices are interactive, multimodal, liberate us from location and at the same time are location-sensitive. What do these mean? Mobile telephony provides the possibility of the "perpetual presence", i.e. we can be reached for different purposes and we can keep in touch with others with no regard to their local conditions and engagement. Mobile devices are capable of connecting people and tasks over vast distances, i.e. the necessary information to the solution of a given problem is accessible due to the accessibility of a wide range of multimodal sources (personal databases, the web, discourse with others, even teleconferencing, and so forth). Since location sensitivity and independence from location are two sides of the same coin, this interactivity could mean help and at the same time some overload. This overload refers to the necessity of multitasking, i.e. it requires quick switching among physical, psychological, and computational conditions.[6] However, multitasking seems to be a certain disturbing concomitant of mobile technology, the high level of connectivity and multimodality could dissolve the load that multitasking induces and help to integrate newly acquired knowledge both on the personal and communal level.

---

[5] Walter J. Ong, *Orality and Literacy: The Technologizing of the Word*, Methuen, London, 1982, p. 136.

[6] Cf. J. S. Rubinstein et al., "Executive Control of Cognitive Processes in Task Switching", in *Journal of Experimental Psychology - Human Perception and Performance*, vol. 27, no. 4, 2001, pp. 763-797. However, the conclusion of the study is that multitasking may actually take more time in the end, i.e. the outcome regarding efficiency is questionable, considering cognitive evolution and the plasticity of the human brain, adaptation to new circumstances is exceedingly possible.

Mauruizo Ferraris considers "[t]he cell phone, bringing together in a single limited physical space all the functions that we have already listed (telephone, radio, cinema, television, e-mail, web and data storage) is the absolute instrument."[7] That is, all the previous inventions of communications technology are integrated into the mobile phone, which means that the mobile device could be the main mediator in social intercourse. Considering the impact of mobile technology from a social and cognitive point of view, I will focus briefly on the integrative potential of multimedia and, after this, on the integrative importance of communication itself.

Multimedia is greatly helpful in gaining new information and getting closer to someone's problem from afar. Just as earlier telecommunication made it possible to hear someone's voice from a great distance, mobile multimedia brings close the experience of the other who is at a distance. If we compare the informative potential and immersive power of images (not to mention videos) and words, it seems obvious that images are the more effective.

The main pillar of the argument for this statement is the special characteristics of visually gained information compared to that which is verbally gained. According to the representational theory of the visual mind, there are two kinds of visual representations:

> visual percepts and visuomotor representations. The former serves as input to higher human cognitive processes, including memory, categorization, conceptual thought and reasoning. The latter is at the service of human action. From the standpoint of our version of the 'two visual systems' hypothesis, vision serves two masters: thinking about, and acting upon, the world.[8]

The inventors of this theory, Pierre Jacob and Marc Jeannerod, claim that although both visual percepts and visuomotor representations are mental representations with non-conceptual content, they still differ significantly from one another.

The authors argue that the content of visual perceptual representations is "both more fine-grained and informationally richer than the conceptual contents of thoughts."[9] Furthermore, this detachment is a necessary consequence of the peculiar situation in which the conceptual transformation of pictorial information means mapping to a considerably narrower range. The visual system provides many kinds of information (e.g., size, orientation, shape, colour, tex-

---

[7] Maurizio Ferraris, "Where Are You? Mobile Ontology", in Kristóf Nyíri (ed.), *Mobile Understanding: The Epistemology of Ubiquitous Communication*, Passagen Verlag, Vienna, 2006, p. 48.

[8] Pierre Jacob and Marc Jeannerod, *Ways of Seeing. The Scope and Limits of Visual Cognition* (2003), Oxford: Oxford University Press, 2004, p. 45.

[9] *Ibid.*, p. 22.

ture, spatial position, distance, motion, etc.) all at once. These data are important either for acting on an object or to define objects' relation to each other, and it is possible to recall them. During the conceptual transformation, we lose some details for the sake of the unambiguousness of verbal expression and in accordance with our priorities. "Now once the visual percept has been turned into a thought by a process involving a selective elimination of information, further conceptual processing can yield a still more complex thought involving, not a two-place relation between pairs of objects, but a three-place relation between a pair of objects and an egocentric perspective."[10] Visuomotor representations are egocentric but in a strictly functional sense: the spatial reference is the actor's body (e.g., if I want to grasp a glass of water the frame of reference is determined by how to capture it, including how far is it from me, what grip size would be convenient, how heavy it might be, etc.). In the case of verbal or conceptual transformation, an egocentric perspective means a conscious relation in accordance with some general goals. As Michael Tomasello put it:

> As perspectivally based cognitive representations, then, linguistic symbols are based not on the recording of direct sensory or motor experiences, as are the cognitive representations of other animal species and human infants, but rather on the ways individuals choose to construe things out of a number of other ways they might have construed them, as embodied in the other available linguistic symbols that they might have chosen, but did not.[11]

This means a certain distance from the surrounding world. To express something verbally means, at the same time, to reorder its elements in accordance with certain goals. "What makes linguistic symbols truly unique from a cognitive point of view is the fact that each symbol embodies a particular perspective on some entity or event: this object is simultaneously a rose, a flower, and a gift."[12] This unique feature of linguistic symbols exists because the socially embedded user has the intention of influencing his/her communicational partners according to his/her priorities. This distanced attitude is not characteristic in the case of other kinds of mental representations. Information mediated by images is much richer in detail and the mental representation of it is either deeply embedded in practice or provides plentiful material of the whole pictorial statement for further processing. Accordingly, images bear a certain immersive potential by which visual information appears as our own experience with greater primacy and without the distanced restructuring power of language.

---

[10] *Ibid.*, p. 31.

[11] Michael Tomasello, *The Cultural Origins of Human Cognition*, Cambridge MA: Harvard University Press, 1999, p. 9.

[12] *Ibid.*, p. 107.

With regards to visuality and mobile technology, several studies emphasize the integrative power of MMS, which partly derives from the fact that it allows us to share our visual experiences continuously and thus create a shared code and practice of interpreting visual experiences. Some authors, including Barbara Scifo, consider the exchange of visual information to be no less than an act of trust,[13] which can help us, among other things, describe and even testify to situations, allowing those situations and experiences to be shared.

A variety of surveys revealed that young people consider sending a pictorial message to be a very convenient, timesaving, and effective means of exchange;[14] moreover, mobile phones capable of storing pictures served as photo albums for their users.[15] Empirical studies[16] confirm that MMS exchanges are characteristic within narrow circles of close friends. This is no coincidence, since the interpretation of a picture requires shared contextual information. As most empirical research emphasizes, pictorial messages are very effective tools for maintaining co-presence and visual presence. Therefore, the mobile phone, (which was already considered a personally tailored device), utilizing visuality as a supplementary tool, can help maintain personal presence and reinforce the contacts of its user.

As we can see, MMS, an aspect of mobile technology that is only a few years old, has considerable integrative potential connected to social intercourse. Images have special immersive potential compared to that of verbal expression. It is easier to integrate new data into the framework of previously acquired knowledge when new information is strengthened by primary visual experiences, and the sharing of visual experiences helps to commit new information to memory. Accordingly, mobile technology, due to its multimodal capability, has considerable integrative potential both in a cognitive and social sense. Moreover, I propose that it is much easier to integrate new elements into the network of the knowledge of the individual if it is possible to record, (or more precisely, to objectify), them through a picture or verbal expression. Placing this image or note into the framework of previously recorded items makes it possible to share it with others at any time. Thus mobile technology, and especially

---

13 Barbara Scifo, "The Domestication of Camera – Phone and MMS Communication", in Kristóf Nyíri (ed.), *A Sense of Place. The Global and the Local in Mobile Communication*, Passagen Verlag, Vienna, 2005, p. 368.

14 *Ibid.*, p. 371.

15 See V. Oksman, "MMS and Its "Early Adopters" in Finland", in Kristóf Nyíri (ed.), *A Sense of Place. The Global and the Local in Mobile Communication*, Passagen Verlag, Vienna, 2005, pp. 349–361.

16 See, for example, N. Döring et al., "Contents, Forms and Functions of Interpersonal Pictorial Messages in Online and Mobile Communication", in Kristóf Nyíri (ed.), *Mobile Understanding: The Epistemology of Ubiquitous Communication*, Passagen Verlag, Vienna, 2006, pp. 197–207.

multimodal capabilities, could provide a visually-based framework of mutual experiences.

According to Dewey's famous formulation:

> Society not only continues to exist *by* transmission, *by* communication, but it may fairly be said to exist *in* transmission, *in* communication. There is more than a verbal tie between the words common, community, and communication. Men live in a community in virtue of the things which they have in common; and communication is the way in which they come to possess things in common. What they must have in common in order to form a community or society are aims, beliefs, aspirations, knowledge – a common understanding – like-mindedness as the sociologists say. Such things cannot be passed physically from one to another, like bricks; they cannot be shared as persons would share a pie by dividing it into physical pieces. The communication which insures participation in a common understanding is one which secures similar emotional and intellectual dispositions – like ways of responding to expectations and requirements.[17]

Some paragraphs later he continues:

> Not only is social life identical with communication, but all communication (and hence all genuine social life) is educative. To be a recipient of a communication is to have an enlarged and changed experience. ... The experience has to be formulated in order to be communicated. To formulate requires getting outside of it, seeing it as another would see it, considering what points of contact it has with the life of another so that it may be got into such form that he can appreciate its meaning.[18]

This shared, (i.e., in some way mediated), experiential background creates the basis of a community. According to Robin Dunbar's *social brain* hypothesis, human intelligence is basically social intelligence; even language originates from the socially determined attitude of humans. Accordingly, gossip (just like grooming in the case of primates) is an extremely important cohesive element in a community. Gossip is possible because the members of the community bear a set of shared information – at the same time gossip maintains and widens this mutual background. According to Dunbar's theory, the size of a group is determined by the size of the members' neocortexes, and mobile technology has no impact on this relationship. However, the flexibility of the connectivity and the immersive potential of new capabilities may frequently reconfigure the circle of people with whom we keep intensive contact.

---

[17] John Dewey, *Democracy and Education. The Middle Works of John Dewey 1899-1924*, vol. 9 (1916), Southern Illinois Univ. Press, Carbondale and Edwardsville, 1985, p. 7.

[18] *Ibid.*, p. 8.

Reordering the Institutional Settings

Casting a glance at the history of education, it becomes clear that it is in close relation to the development of the concept of childhood, the alteration of family life, and intellectual attitudes in general. From the late 16th and particularly from the 17th century, a slight and partial, albeit important, change emerged in family life. Until the 17th century, life meant public life. "[In] fact privacy scarcely ever existed, when people lived on top of one another, masters and servants, children and adults, in houses open to all hours to the indiscretions of callers".[19] This did not mean that family did not exist. Rather, it meant that the private and public spheres were not separated as sharply as we are accustomed. According to Ariès' research, the isolation of the individual and family significantly emerged in paintings from the 17th century. Earlier, "the important thing was the representation of public life. This representation doubtless corresponds to a profound reality."[20]

Similar changes are noticeable in the concept of childhood and education. The amusing fact that in medieval society the idea of childhood, (i.e. "an awareness of the particular nature of childhood"), did not exist,[21] means at the same time that "[society] had no idea of education" as well.[22] In the Middle Ages, school was not considered an institution to educate children. The intention to educate mostly youth in so-called colleges only emerged at the beginning of modern times. The other "essential difference between medieval school and the modern college lies in the introduction of discipline."[23] Discipline means better supervision inside the school and at the same time the expectation that the boundaries of the school period will be respected. This change emerged in the late 16th and 17th century.

"But why did the school system itself emerge?" asks Kristóf Nyíri. "The explanation is obvious, even though Ariès fails to see it. As Neil Postman, combining Ariès with McLuhan, points out: it was the rise of *literacy* that made formal schooling inevitable."[24] The influence of literacy was also significant in the realm of philosophy (think of the Cartesian ego), and it is from the 17th century onward that we can talk about science. The influence of literacy and the

---

[19] P. Aries, *Centuries of Childhood: A Social History of Family Life*, transl. from the French by Robert Baldick, Random House, New York, 1962, p. 405.

[20] *Ibid.*

[21] *Ibid.*, p. 128.

[22] *Ibid.*, p. 411.

[23] *Ibid.*, p. 333.

[24] Kristóf Nyíri, "Towards a Philosophy of M-Learning", in M. Milrad – U. Hoppe – Kinshuk (eds.), *Wireless and Mobile Technologies in Education*, CA: IEEE Computer Society, Los Alamitos, 2002, p. 122.

printing press had by that time reached quite a high level, i.e. its influence was significant in many fields of life, including the Law, the accumulation of knowledge, conceptual abstractness, and analysing attitudes among educated members of society. The possibility of differentiated legislation makes more complicated organisational forms possible and, on the other hand, a distanced analytical attitude was able create the demand to conceptually and institutionally different-tiate between child and adult.

> The first concept of childhood – characterised by 'coddling' – had made its appearance in the family circle, in the company of little children. The second, on the contrary, sprang from a source outside the family: churchmen or gentlemen of the robe, few in number before the sixteenth century, and a far greater number of moralists in the seventeenth century, eager to ensure disciplined, rational manner.[25]

By the 19th century, the demarcation of private and public life, and the educational system, had crystallized. However, criticism of it emerged in the early 20th century, from Dewey.

Dewey considers education to be deeply embedded into everyday communication and activity. He creates the "concept of learning through experience, the concept of learning by engaging in activities", as Seymour Papert describes Dewey's attitude. During the centuries of literacy, institutions were tailored according to clear-cut alphabetical reason, which means a theoretically and verbally based amount of knowledge to be learnt in a sequential order separated from other fields of life. "We try in our school systems to decide what people will learn in this top-down, centralized way and, for the same reason, it is not compatible with the complexities and dynamic possibilities of the modern world", as Papert puts it.[26] These characteristics of the educational system dominate our culture. According to Ivan Illich

> everywhere the school system has the same structure, and everywhere its hidden curriculum has the same effect… This identity of school system forces us to recognize the profound world-wide identity of myth, mode of production, and method of social control, despite the great variety of mythologies in which the myth finds expression.[27]

That is, the influence of literacy is recognisable everywhere in the world. Since education evolved according to the limits and demands of literacy, it fits our present conditions poorly. Institutional practice tries to integrate new technology, but at the same time it deprives it of its original framework, which en-

[25] P. Aries, *Centuries of Childhood*, p. 132.

[26] Seymour Papert, *Child Power: Keys to the New Learning of the Digital Century* (1998) http://www.connectedfamily.com/main_alt.html , last accessed Febr. 26, 2007.

[27] Ivan Illich, *Deschooling Society* (1971), London: Marion Boyars, 1978, p. 74.

tails a certain distortion in the ordinary manner of usage. Let me quote Seymour Papert again:

> instead of the computer cutting across the disciplines in the subjects, the computer is now confined to a computer room and it is a subject of its own, taken out of the mainstream of the learning environment. There's now a specialized computer teacher. There's a curriculum even for the study of the computer. It has been normalized by the system; it has been tamed. That's not the only way in which it can be tamed.[28]

Criticism of the educational system mostly touches upon the conservative character of education, which means a deep-rooted commitment to literacy. Dewey recognized that the accustomed role of education is based on traditional dualisms (like matter vs. method, intellect vs. emotion, activity vs. passivity, empirical vs. higher rational knowing, etc.), with the focus only on one of these pairs at a time. "All of these separations culminate in one between knowing and doing, theory and practice, between mind as the end and spirit of action and the body as its organ and means."[29] This comprehension of human activity is one of the most influential findings of the literate mind. Criticism of this dualistic attitude became very active in 20th century philosophy. Dewey connected it with science and education by the formulation of a demand for a new theory of knowledge

> which sees in knowledge the method by which one experience is made available in giving direction and meaning to another. The recent advances in physiology, biology, and the logic of the experimental sciences supply the specific intellectual instrumentalities demanded to work out and formulate such a theory. Their educational equivalent is the connection of the acquisition of knowledge in the schools with activities, or occupations, carried on in a medium of associated life.[30]

Papert describes this situation as the dominance of the "informational side" compared with "doing things, making things, [and] constructing things". The top-down model of education and the emphasis on the "informational side" of knowledge are equal consequences of literacy, since its ideals suggest that there is a necessary amount of objective knowledge which can be learned as a verbally expressed record. The teacher, as the living and authentic source of this knowledge, mediates the necessary information. Illich was right in thinking that it is not possible to go beyond recent conditions "unless we first understand that obligatory public schools inevitably reproduce such a society, no matter what is taught in them."[31] Since education passes down patterns which are rooted in the

---

[28] Papert, *Child Power*.

[29] Dewey, *Democracy and Education*, p. 346.

[30] *Ibid.*, pp. 354 f.

[31] Illich, *Deschooling Society*, p. 38.

past, there is not much hope for change. This is the reason why changes in the educational system have been so slow and have disregarded criticism for nearly a hundred years.

> A good educational system should have three purposes: it should provide all who want to learn with access to available resources at any time in their life; empower all who want to share what they know to find those who want to learn it from them; and, finally, furnish all who want to present an issue to the public with the opportunity to make their challenge known.[32]

Instead of a top-down model, he suggests creating networks. With such networks, it would be easy to answer new challenges. The rapid change of technology and, accordingly, the framework of everyday activity require quick adaptation. Thus, it has recently become true that "[t]he one really competitive skill is the skill of being able to learn."[33]

Illich's suggestion is quite timely in the age of networks. New information technology, especially mobile technology, is particularly capable of serving as the basis for a flexible interchange of different kinds of knowledge and expertise, as well as of mediating among different kinds of networks. One can object that some basic skills must be formally learnt within the traditional framework, otherwise there is no capability of taking part in this interchange. This objection is easy to ward off either on the basis of Dewey's educative communication theory, or in light of recent rather convincing experiences involving the educative power of motivational possibilities such as keeping in touch via chatting, mutual interest and so forth.

Conclusion

Since mobile technology unifies almost all the mediators of life, (there are different inter-mediators of representations which make them accessible to others), it appears to be very successful and likely to become the dominant medium. This new kind of medium bears considerable integrative potential both from a cognitive and a social point of view. It creates a new cohesive foundation among people and also makes a flexible reordering of close relations possible. Regarding mobile learning, these capabilities will be considered within the framework of a conservative institutional structure of education which is committed to the demands of literacy. Schools preserve and maintain the cultural patterns which created them. This kind of reproduction of the traditional patterns makes change very difficult.

---

[32] *Ibid.*, p. 75.

[33] Papert, *Child Power.*

Preserving knowledge and passing it on have always been crucial issues for any community. If we consider the history of the institution of education, (and that of the museum or library), it becomes clear that not only the Law, but the entire institutional framework of life has been influenced by, and reshaped according to, the literate mind for hundreds of years. According to the nature of technological inventions, they are designed to fit recent demands and thus their influence over the long run is hardly tangible at the moment. From the end of the 19th century, inventions of communications technology began to burst the traditional conceptual and even perceptual framework of the literate mind. (Think of the overcoming of distance and time by telecommunications, sound and image-recording technologies, etc.)

Allow me to quote Ron Burnett to indicate the significance of recent changes: "The capacity to watch and experience the news from everywhere and the 'idea' that time (and different contexts) can be shared contributes to the intuition that these events are not being lived within conventional definitions of time and space."[34] These two issues (time and space) are decisive in the case of mobile technology. Apart from strictly conceptual considerations, Dewey's emphasis on communication and Illich's idea regarding the re-configuration of education along networks fit the capacities of current technology very well. That is, mobile devices create convenient educational circumstances based merely on their prime capabilities. Since these means are quite ready at hand, their use does not mean any extra load on cognitive capacities. Moreover, they yield considerable help in getting used to changing circumstances. Papert calls attention that taming new technology may not exclusively mean that it becomes integrated by breaking it off from its ordinary use. Mobile devices are part of everyday life, they serve their intended purposes and at the same time are very effective at helping education in Dewey's sense. Mobile learning could provide an outline for an educational system which is deeply immersed in everyday practice, yields easy access to information for different purposes, and could create extensive and flexible networks of individuals.

[34] R. Burnett, *How Images Think*, Massachusetts and London : MIT Press, Cambridge, 2004, p. 73.

LANGUAGE AND COGNITION

Old Patterns, New Bewitchments*

Recent inventions in communications technology have given rise to several changes in the framework of everyday activities. These changes herald an imminent transformation in the institutional framework, too. In the present paper, I'd like to focus on the conceptual background of some obstacles which slow this transformation down. Learning in the context of mobile communication (m-learning), a recent and growing research field, provides a good example of the decisive influence of old patterns while at the same time it emphasizes the possibilities opened up by new communications technology. This perspective offers a good opportunity to connect philosophical considerations, especially Wittgenstein's, to practical issues, and it also facilitates seeing Wittgenstein's efforts as a criticism of these hindering patterns. Some of Wittgenstein's more opaque objections can today be better understood in the light of recent cognitive science.

New Conditions, Old Institutions

Education forms an important segment of social life and at the same time mirrors the dominant views on knowledge and the general order of things. John Dewey called attention to exactly those traditional dualisms which create the background of education and emphasized the importance of an awareness of these mostly non-reflected presuppositions. Even though Dewey outlined his view at the start of the 20th century, the situation has not really changed.

There is a deep interrelatedness between Dewey's criticism and changes in communications technology. As Kristóf Nyíri puts the matter: "it was the rise of *literacy* that made formal schooling inevitable"[1]; at the time when Dewey's criticism was formulated, the outlines of the shift to the age of *secondary orality* were emerging. According to Walter J. Ong, the institutional and also the cognitive framework changed as a consequence of the invention of alphabetical writing. Alphabetical writing is a very effective means of storage; however, there is no supporting environment or unity of references that facilitate understanding. As a new kind of storage system with built-in barriers, alphabetical writing, as compared to orality, created new habits in the thought process.

---

* First published in H. Hrachovec, A. Pichler, and J. Wang (eds.), *Philosophy of the Information Society. Papers of the 30th International Wittgenstein Symposium*, 2007, pp. 109-111.

[1] Kristóf Nyíri, "Towards a Philosophy of M-Learning" in M. Milrad, U. Hoppe and Kinshuk, (eds.) *Wireless and Mobile Technologies in Education*, Los Alamitos: CA: IEEE Computer, 2002, p. 122.

General subjects, abstract concepts detached from the human life-world, and linearly-structured arguments emerged.[2] The cohesive power of live intercourse and its multimodal contexts remain an important factor in everyday life, but the conceptual and institutional frameworks have gradually been altered by the requirements of silent texts. In the age of secondary orality, there are other mediators of experience and ideas beside texts. The emergence of different recording technologies and systems for communicating over distance gradually allowed, once more, more direct access to others' experiences. Accordingly, the pressure of verbal formulation decreased, and mediated communication gradually became closer to live intercourse.

Recently, we can see that the traditional differentiations of work and spare-time, the public and private sphere, learning and post-learning, workplace and home, theory and practice, etc. seem to have disappeared from our everyday routine. Teleworking, lifelong learning, and the continuous rescheduling of tasks are now part of everyday activities. These changes indicate an overall transformation in the social, institutional, and even cognitive framework. However, if we cast a glance at researchers in the field of education, we find that they mostly move within the traditional framework both in a conceptual and an institutional sense. That is, they mostly try to tame new technology to fit into the institutional setting, divesting it of its original use. Of course, there are efforts to incorporate everyday practice, such as introducing more fieldwork and teamwork, and utilizing multimodal means to help the learner, but the gap between school and everyday life (in contrast to the above-mentioned tendencies) seems to have been maintained.

Subsequently, I will focus on the roots of this state of affairs. I rely on Dewey's considerations and relate them to Wittgenstein's famous phrase: "Philosophy is a battle against the bewitchment of our intelligence by means of language".[3]

Old Patterns and Looking for New Ways

I suggest examining the above phrase as one which errs to some extent, and at the same time is deeply correct. At first sight the objection arises that it is not language, but our view of language, that is responsible for the bewitchment. However, considering recent studies in the realm of cognitive psychology, it becomes clear that some characteristics of language, especially the circumstances of literacy, indeed make it capable of leading us astray as regards our views on the nature of language.

---

[2] See Walter J. Ong, *Orality and Literacy: The Technologizing of the Word*, London: Methuen Ong 1982, especially pp. 31-57 ff. and 103-112 ff.

[3] Ludwig Wittgenstein, *Philosophical Investigations*, Oxford: Basil Blackwell, 1963, § 109.

These characteristics of language become especially visible if we compare the cognitive processing of verbal and pictorial representations. According to Allan Paivio's dual coding approach, imagery and verbal expression are parallel processes, but some important differences between them are discernable. As Paivio puts it,

> verbal descriptions of concrete situations and events from memory and verbal expressions of the manipulation of spatial concepts are likely to be mediated efficiently by non-verbal imagery, whereas abstract discourse and verbal expressions of abstract reasoning are more likely to be mediated entirely by the verbal system. A second (less obvious) implication is that the verbal behavior mediated by imagery is likely to be more flexible and creative than that mediated by the verbal symbolic system. This follows from the theoretical assumption that the spatially and operationally parallel image system is not characterized by logical sequential constraints to the same degree as the verbal symbolic system.[4]

In Jacob and Jeannerod's representational theory of visual mind, neither of the two kinds of visual representations (visual percepts and visuomotor representations) have conceptual content, yet both of them can serve as the basis for or be subject to conceptual processing. The visual system provides many kinds of information all at once. These data are important either for acting on an object or for defining objects' relations to each other, and it is possible to recall them. During the conceptual transformation, we lose some details, resulting in an increase in the unambiguousness of verbal expression and in accordance with our priorities.

> Now once the visual percept has been turned into a thought by a process involving a selective elimination of information, further conceptual processing can yield a still more complex thought involving, not a two-place relation between pairs of objects, but a three-place relation between a pair of objects and an egocentric perspective.[5]

Visuomotor representations are egocentric, but in a strictly functional sense: the spatial reference is the actor's body. In the case of verbal or conceptual transformation, an egocentric perspective means a reflexive relation in accordance with some general goals. According to Michael Tomasello:

> As perspectivally based cognitive representations, then, linguistic symbols are based not on the recording of direct sensory or motor experiences, as are the cognitive representations of other animal species and human infants, but rather on the ways individuals choose to construe things out of a number of other ways they might have construed

[4] Allan Paivio, *Imagery and Verbal Processes* (1971), Hillsdale NJ: Lawrence Erlbaum Associates, Publishers, 1979, pp. 434 f.

[5] Pierre Jacob and Marc Jeannerod, *Ways of Seeing. The Scope and Limits of Visual Cognition* (2003), Oxford: Oxford University Press, 2004, p. 31.

them, as embodied in the other available linguistic symbols that they might have chosen, but did not.[6]

This entails a certain distance from the surrounding world. To express something verbally means, at the same time, to reorder its elements in accordance with certain priorities. This unique feature of linguistic symbols exists because the socially embedded user has the intention of influencing his/her communicational partners according to his/her priorities.[7] This distanced attitude does not characterize other kinds of mental representations.

Summing up these characteristics, we can claim that verbal expression (compared to other kinds of representation) strengthens the impression that thinking is a process bound to the individual related to any kind of bodily activity only as far as certain goals which are anchored in practice serve as the basis of it. Literacy and its institutions (like silent reading, libraries, school discipline, etc.) intensify this impression and lead to the conviction that intellectual engagements are solely activities separated from practice. Dewey recognized that the customary role of education is based on traditional dualisms (such as matter vs. method, intellect vs. emotion, activity vs. passivity, particular vs. universal, empirical vs. higher rational knowing, etc.), with the focus only on one of these pairs at a time. "All of these separations culminate in one between knowing and doing, theory and practice, between mind as the end and spirit of action and the body as its organ and means."[8] This comprehension of human activity is one of the most influential ideas of the literate mind. Criticism of this dualistic attitude became very active in 20th century philosophy. Dewey himself offers a different view relying on the findings of physiology and psychology; he suggests replacing the old dualism of body and soul with that of brain and the rest of the body, and to further regard it as a whole where "brain is the machinery for a constant reorganizing of activity".[9]

There are the sounds of the words, and all sorts of bodily sensations connected with gesture and intonation. Where we are liable to go wrong is in supposing that sensations connected with words are somehow 'in the mind'.[10]

---

[6] Michael Tomasello, *The Cultural Origins of Human Cognition* Cambridge, MA: Harvard University Press, 1999, p. 9.

[7] For the importance of social embeddedness, see Robin Dunbar's theory of "social intelligence" and Merlin Donald's reconstruction of cognitive evolution.

[8] John Dewey, *Democracy and Education. The Middle Works of John Dewey 1899-1924*, vol. 9, Carbondale and Edwardsville: Southern Illinois University Press, 1985, p. 346.

[9] *Ibid.*

[10] Ludwig Wittgenstein, *Wittgenstein's Lectures. Cambridge, 1932-1935*, Oxford: Basil Blackwell, 1979, p. 114.

Wittgenstein claims that this kind of error is, to some extent, related to language. However, some of his remarks suggest that the language that we use is optimal because we gain the most appropriate descriptions through it. His remarks on understanding, ideas, the use of symbols and rules lead to the statement: "What we are apt to confuse is the idea as a state of mind occurring at a particular time and the use we make of that idea."[11] It is as if Wittgenstein believed that we are inclined to emphasize the thing-like/substantive and static character of a given phenomena, whereas the active component of the state of affairs is concealed. He claims that "[o]ne of the chief troubles is that we take a substantive to correspond to a thing. Ordinary grammar does not forbid our using a substantive as though it stood for a physical body."[12] And accordingly, it is easy to mix up the rules we use, i.e. we are inclined to forget that not all substantives are things. However, as he suggests, we "desire to point to something". This touches upon the metaphoric nature of language, or more precisely, that of our thought. As Lakoff formulates it, "the locus of metaphor is not in language at all, but in the way we conceptualize one mental domain in terms of another... [M]etaphor (that is, cross-domain mapping) is absolutely central to ordinary natural language semantics".[13] The cognitive approach to metaphors clearly points out that even abstract concepts (such as time, states, change, causation, etc.) have their roots in everyday activity.

Wittgenstein's tentative suggestions regarding prejudices in relation to the nature of language have been corroborated, since there is empirical and theoretical evidence by which we can argue for them, at least from a cognitive point of view. Paivio notes that Berlyne "accepts the usefulness of words as situational (that is labeling) responses, rather surprisingly argues that verbal processes are deficient in their capacity to represent transformations".[14] This is a disputable question for Paivio because he finds it difficult to prove that the motor component is more intrinsic to images. I believe that the representational theory of visual mind yields a clarifying distinction: the distinction between visual percept which "serves as input to higher human cognitive processes, including memory, categorization, conceptual thought and reasoning" and visuomotor representation which "is at the service of human action".[15] Both can be at the service of action, but acting upon the world without visuomotor representation is at the very least impaired and cumbersome, whereas conceptual

---

[11] *Ibid.*, p. 87. See also: "The phrase 'in the mind' has caused more confusion than almost any other in philosophy" (*ibid.*, p. 114).

[12] *Ibid.*, pp. 31 f.

[13] John Lakoff, "The Contemporary Theory of Metaphor" in Andrew Ortony (ed.), *Metaphor and Thought*, Cambridge University Press, 1993, p. 203.

[14] Paivio, *Imagery and Verbal Processes*, p. 31.

[15] Jacob and Jeannerod, *op. cit.*, p. 45.

processing of visually-gained information doesn't necessarily need a visuomotor supplement. That is, visuality/imagery is closer to action than conceptual processing. At a primary level, the relationship between them is without mediation. This, of course, does not mean that verbally processed and stored information has no role in acting, it just draws attention to the notion that there is a difference in grade.

Conclusion

As we can see, some of the difficulties Wittgenstein touched upon have continued to surface in a considerable body of recent research (in cognitive and developmental psychology, as well as in the contemporary theory of metaphors). Although these difficulties were encountered somewhat prior to Wittgenstein (Bergson, James, and Dewey come to mind), the earlier findings didn't have the impact on philosophical thinking that they deserved. The exceptional contributions of Bergson and Dewey originate from their specific interests: Bergson studied the scientific results of his time, while education, a practical issue, was one of Dewey's chief concerns. This deeper embeddedness in practical and scientific problems could help distance us from the dominant paradigm of verbal expression. As Wittgenstein indicates, the dominance of verbal expression can indeed distort experience to some extent. Recently, means have become available to mediate experience and thought not only verbally, but multi-modally; that which is mediated is close to, or even identical with, experience. We are no longer as much at the mercy of verbal expression as we were. Accordingly, past theoretical considerations can be examined in a new light. The feeling of mental discomfort can be interpreted as the awareness of certain dissonances which originate from the chasm between experience on the one hand, and theories that describe reality on the other. Even though "Wittgenstein's later work can be usefully interpreted as a philosophy of post-literacy",[16] Wittgenstein lacked the proper tools, within the framework of the traditionally elaborated conceptual network of literacy, to grasp the relevant new experiences.

[16] Kristóf Nyíri, "Wittgenstein's Philosophy of Pictures", in Alois Pichler – Simo Säätelä (eds.), *Wittgenstein: The Philosopher and his Works*, Working Papers from the Wittgenstein Archives at the University of Bergen, no. 17, 2005, p. 352. For more details see Kristóf Nyíri, "Wittgenstein as a Philosopher of Secondary Orality" in: *Grazer Philosophische Studien* 52 (1996/97), pp. 45-57.

Contexts: Basic and Metaphoric*

In recent literature on the problem of cognition, knowledge, and human activity, the notion of 'context' has achieved paramount significance. In this paper, I argue for a much more radical view: I suggest that from both a cognitive and a metaphysical point of view, the separation of the human mind and its external environment paralyzes the interaction between them, and also the human mind itself. To prove my hypothesis, I first relate this distinction to the objectivist tradition. Then, relying on cognitive psychological research, I try to show the importance of context both on a basic and a conceptual level. Finally, I point out the interrelatedness of the basic and conceptual levels. In doing this, I rely considerably on Henri Bergson's attempt to unify body/spirit dualism and the metaphor theory of George Lakoff.

Context Revisited

What does 'context' mean? The Latin root *contexere* means to intertwine, to twist, join, and weave together. More recent usage refers to a word or passage's surroundings that can throw light on its meaning, as well as to a certain environment, specifically "the interrelated conditions in which something exists or occurs".[1] We can talk about the context *of something,* i.e. context requires the presupposition of the distinction of essential and non-essential parts of a whole. This distinction is closely related to the objectivist point of view, and obviously, to the objectivist theory of categorization.

According to George Lakoff, objectivism presupposes that:

> All of reality consists of entities, which have fixed properties and relations among them at any instant… Among the properties that things have, some are essential; that is, they are those properties that make the thing what it is and without which it would not be that *kind* of a thing. Other properties are accidental – that is, they are properties that things happen to have, not properties that capture the essence of the thing.[2]

---

* This is a revised version of a talk given in June 2007 at the London Knowledge Lab, at the conference *Knowledge and Context.*

[1] Webster's Dictionary, http://www.merriam-webster.com/dictionary/context

[2] George Lakoff, *Women, Fire, and Dangerous Things. What Categories Reveal about the Mind,* Chicago and London: The University of Chicago Press, (1987) 1990 pp. 160. See also: "The essential thing that this is a *lamp,* that it serves to give light; – that it is an ornament to the room, fills an empty space, etc., is not essential. But there is not always a sharp distinction between essential and inessential." (Ludwig Wittgenstein, *Philosophical Investigations,* Oxford: Basil Blackwell 1963 p. 30.) This remark of Wittgenstein touches upon the problem of

The concept of *category* has a crucial role in the objectivist account of reality. Classical categories are "defined by the properties common to all their members".[3] According to this model, the world is "extremely well-behaved" because:

> It is made up of discrete entities with discrete logical combinations of atomic properties and relations holding among those entities. Some properties are essential; others are accidental. Properties define categories, and categories defined by essential properties correspond to the kinds of things that there are. And the existence of classical categories provides logical relations that hold objectively in the world.[4]

There is a firm basis for the long-lasting success of the classical theory of categorization, which is reinforced by clearly visible implications in mathematics and computer science.[5]

According to the objectivist comprehension of reality, there is no interaction between human understanding and the reality which is to be understood. As Lakoff puts it: "The objectivist paradigm can be broken down into two parts: [m]etaphysics [and] [e]pistemology".[6] This means that human understanding is considered an independent phenomenon without any relation to its circumstances. This is possible because of some additional presuppositions, *viz.* that the human mind works as a machine which manipulates symbols. These symbols represent the entities discussed above, as well as the properties of the entities and their relations. Symbols are "internal representations of external reality" and gain their meaning "via correspondence to the things in the external world".[7] Thought is characterized by a certain atomism, i.e. thoughts are made up of simple "building blocks". These atomistic blocks can be manipulated in accordance with general principles.

Ernest Gellner labeled this tradition "atomistic individualism". Gellner uses the attribute "atomistic" to emphasize the individual character of the epistemology of this tradition, with Descartes as the obvious lynchpin.

> Descartes was perhaps the chief, certainly the most famous and elegant progenitor of intellectual individualism ...He insisted that true knowledge could best be obtained by a single individual, who had bravely and ruthlessly freed himself from the incubus of the conventional wisdom of his own culture and had built up a new capital exclusively from neat, distinct, clear elements, separate from each other. Acting alone, step by separate

---

categorization and converges with Lakoff's considerations, viz. that objectivist categorization is not capable of describing the everyday/real use of categories.

[3] Lakoff, ***Women, Fire, and Dangerous Things***, p. xiv.

[4] *Ibid.* p. 162.

[5] *Ibid.* p. 184.

[6] *Ibid.* p. 159.

[7] *Ibid.* p. xii.

step, *that* is the basic rule of the procedure... He trades only with his own self-made capital and need fear no taint which might devalue his future achievements.[8]

The Cartesian distinction of *res cogitans* and *res extensa* sets a certain distance from the human body. Since the body belongs to the world of extended things, the traditional dualism of the mind (or soul, or spirit) and the body reaches its most radical stance. A thinking entity has no direct access to the extended body and, accordingly, the gap between body and mind becomes unbridgeable. To this challenging paradox was born some solutions which highlighted exactly the importance of context.

With regard to chronology, I should first mention Henri Bergson. He describes the situation as follows:

> The mistake of ordinary dualism is that it starts from the spatial point of view: it puts, on the one hand, matter with its modifications, in space; on the other hand, it places unextended sensations in consciousness. Hence the impossibility of understanding how the spirit acts upon the body or the body upon the spirit. ... But hence also the impossibility of constituting either a psychology of memory or a metaphysic of matter.[9]

Bergson's solution rests upon the close interrelatedness of the external and internal via perception, motor activity, and memory. As he writes, "[t]he reality of matter consists in the totality of its elements and their actions of every kind. Our representation of matter is the measure of our possible action upon bodies."[10]

Bergson emphasizes the material character both of the things of the external world and the human body. At the same time, he stresses the mutual interrelatedness of material entities and the special character of the human body "which acts like other images[11], receiving and giving back movement, with, perhaps, this difference only, that [my] body appears to choose, within certain limits, the manner in which it shall restore what it receives".[12] This stands in contrast with the traditional conviction which considers "the grey matter and its modifications ... as things which are sufficient to themselves and might be

---

[8] Ernest Gellner, *Language and Solitude. Wittgenstein, Malinowski and the Habsburg Dilemma*, Cambridge: University Press 1998, pp. 7. Gellner stresses the role of Hume, since in his case the "independence of atomic elements" is sharply highlighted. "Nothing was inherently linked to anything else, the base-line of knowledge was an assembly of disconnected atoms." (*Ibid.* p. 8.) See also pp. 181.

[9] Henri Bergson, *Matter and Memory*, New York: Zone Books (first published in French, 1896), 1991. pp. 220 f.

[10] *Ibid.*, p. 38.

[11] "*I call* matter the *aggregate of images, and* perception of matter *these same images referred to the eventual action of one particular image, my body.*" (*Ibid.*, p. 22.)

[12] *Ibid.*, p. 19.

isolated from the rest of the universe", and "the nervous system as a separate being" [13].

Thus, we can say Bergson tries to unify the traditional spirit/body and at his time accustomed body/brain division. According to Bergson, humans are committed to the material world in a threefold sense: body, and brain too, are made of matter and to perceive the material world is possible via these material entities. Motor activity is bound to the material world, and finally, memory that helps orientation is the recollection of past material experiences. It is memory where Bergson finds "[w]e can grasp spirit in its most tangible form".[14]

That is, Bergson argues for the possibility of the psychology of memory and the metaphysics of matter and, moreover, he tries to prove these are bound up with each other.[15] To formulate this in Lakoffian terms, Bergson intends to unify the detached epistemology and metaphysics of the objectivist tradition.

John Dewey, relying on the scientific results of his time, argues against the separating and dualistic approach. As he writes: "All of these separations culminate in one between knowing and doing, theory and practice, between mind as the end and spirit of action and the body as its organ and means."[16] Relying on the findings of physiology and psychology, he concludes that

> [n]o one who has realized the full force of the facts of the connection of knowing with the nervous system and of the nervous system with the readjusting of activity continuously to meet new conditions, will doubt that knowing has to do with reorganizing activity, instead of being something isolated from all activity, complete on its own account.[17]

---

[13] *Ibid.*, p. 24 Cf. "We are too much inclined to regard body as a world within a world, the nervous system as a separate being, of which the function is, first, to elaborate perceptions, and then, to create movements. The truth is that my nervous system, interposed between the objects which affect my body and those which I can influence, is a mere conductor, transmitting, sending back or inhibiting movement." (*Ibid.* p. 45.)

[14] *Ibid.* p. 73. See also: "Our daily life is spent among objects whose very presence invites us to play a part: in this the familiarity of their aspects consists. Motor tendencies would, then, be enough by themselves to give us the feeling of recognition. But we hasten to add that in most cases there is something else besides.
...our past physical life is there." (*Ibid.* p. 95.)

[15] "We have striven to show that this psychology [psychology of memory] and this metaphysic [metaphysic of matter] are bound up with each other and that the difficulties are less formidable in a dualism which, starting from *pure* perception, where subject and object coincide, follows the development of the two terms in their respective durations..." (*Ibid.* p. 221.)

[16] J. Dewey, *Democracy and Education. The Middle Works of John Dewey 1899-1924,* vol. 9, Carbondale and Edwardsville: Southern Illinois Univ. Press 1985, p. 346

[17] *Ibid.*, p. 347.

Considering the results of biology and contemplating the "doctrine of evolution", Dewey claims that "the living creature is a part of the world", which implies a new comprehension of knowledge. "If the living, experiencing being is an intimate participant in the activities in the world to which it belongs, then knowledge is a mode of participation, valuable in the degree in which it is effective. It cannot be the idle view of an unconcerned spectator."[18]

As we can see, both Bergson and Dewey are convinced that the objectivist account of knowledge and reality meets neither the results of science nor the requirements of a coherent metaphysics.

Martin Heidegger's main endeavour was the "destruction" of traditional metaphysics. His starting point was to highlight the embeddedness of human beings. *Being-in-the-world* is a key term in his work. The surrounding world emerges as a totality of references to help the orientation of acting humans, as opposed to the analyzing, alienated attitude of the investigating subject in accordance with his predecessors' accounts. Heidegger introduced the terms of *ready-at-hand* and *present-at hand* to capture the difference between the perspective of being engaged in and acting upon the world, and that of the observing attitude that was previously the only characteristic of humans worthy of attention for most philosophers.

Basic Context

A great deal of recent research provides information about the cognitive aspects of being engaged in the world. Evolutionary surveys, developmental psychology, and cognitive psychology furnish evidence that cognitive capacity is in close relation to, (and even, in a certain sense, dependant upon), everyday mundane non-reflected activity.

Jerome S. Bruner differentiates three kinds of internal representations: action, i.e. motor activity; imagery; and symbolic, i.e. language-like representations. He emphasizes that "the three representational systems are parallel and each is unique, but all are also capable of partial translation, one into the other."[19] Nevertheless, motor activity plays a fundamental role in the cognitive development of the individual.[20]

Merlin Donald's theory of cognitive evolution relies heavily on the social intelligence hypothesis of Robin Dunbar,[21] which suggests a deep embeddedness

[18] *Ibid.*

[19] Jerome S. Bruner, Rose R. Oliver and Patricia M. Greenfield, et al., *Studies in Cognitive Growth* (1966), New York/London/Sydney: John Wiley & Sons, Inc. 1967, p. 11.

[20] See Bruner et al, *op. cit.*, pp. 16.

[21] See in chapter "Multimodal Integration. From a Philosophical Point of View", in the present volume pp. 124 f.

in everyday activity. Donald's reconstruction of cognitive evolution is built upon the evolution of representative capabilities, and each new representational mode encapsulates the former.[22] Episodic and mimetic representations are closely related to motor activities, and mimetic representation is considerably effective in socially determined emotions, roles, and different motor skills (e.g. the techniques of various crafts). That is, motor activity plays an important role in cognitive functioning, even in the case of modern humans.

Embodiment is a key term in Lakoff's theory of categorization, and it is also crucially important to the question of context. It means that "conceptual structure is meaningful because it is embodied, that is, it arises from and is tied to, our preconceptual bodily experiences. In short, conceptual structure exists and is understood because preconceptual structures exist and are understood."[23] Of course, some questions arise, such as what kind of preconceptual structure could determine the structure of conceptual structure, and how could abstract reason arise from bodily experiences?

According to Lakoff, the preconceptual level is determined by kinesthetic image schematic structures and basic-level categories. Basic-level categories emerge quite early (at the age of three) during ontogenesis. They do not overload memory capacity and are closely bound to motor activity. Neither the subordinate nor the superordinate-level categories can satisfy both requirements. (Consider, for example, the basic-level category of 'chair'. It has many subordinate categories, i.e. there are great many kinds of chairs, which means a considerable load on memory. And to its superordinate category, 'furniture', there is no chance to attach one special motor action.)[24] Here, basic-level does not imply any kind of primitiveness or lack of structure. On the contrary, basic-level categories emerge as gestalt, i.e. they emerge in overall general forms and at the same time their structures are identifiable as well. Since basic-level categories are not simple or atomistic, but they instead bear a certain structure, they do not work as building blocks as supposed by objectivist theory.

Regarding kinesthetic image schemas, Lakoff refers to Mark Johnson's basic insight that

---

[22] "Mimetic representations are metaphoric, rehearsable self-representations built upon episodic knowledge; therefore, by definition, episodic outputs are available to the mimetic representational process. ...This relationship is asymmetrical: episodic systems cannot model the outputs of the mimetic representational system." After the emergence of language, mimetic representations "suffered the same fate as episodic experience, they could not model linguistic content. Nevertheless, mimesis continues to exist as an independent representational system ... But a linguistically able mind will relegate mimesis to a secondary role". (Merlin Donald, *Origins of the Modern Mind. Three Stages in the Evolution of Culture and Cognition*, Cambridge: Harvard Univ. Press, 1993, p. 271.)

[23] Lakoff, *Women, Fire, and Dangerous Things,* p. 267.

[24] For a short introductory summary of basic-level categories, see *ibid.*, p. 46.

> [E]xperience is structured in a significant way prior to and independent of, any concepts. Existing concepts may impose further structuring on what we experience, but basic experiential structures are present regardless of any such imposition of concepts. This may sound mysterious, but it is actually very simple and obvious, so much so that it is not usually considered worthy of notice.[25]

Kinesthetic image schemas originate from everyday bodily experiences and mean certain, relatively simple spatial relations such as: containers, paths (movement in space with a starting and ending point including direction), links (regarding security and its source), forces, balance, different kinds of orientation (up-down, front-back, part-whole, center-periphery, etc.)[26]

Metaphoric Context

Reaching abstract reason from the preconceptual level is possible due to the "conceptualizing capacity" of humans. This capacity consists of:

> - The ability to form symbolic structures that correlate with preconceptual structures in our everyday experience. Such symbolic structures are basic-level and image-schematic concepts.[27]

Let me note that we can speak about basic-level categories in a twofold sense: categories that emerge out of the network of motor activities and categories that are in accordance with the former and refer to concepts in linguistic form. Our conceptualizing capacity also consists of:

> - The ability to project metaphorically from structures in the physical domain to structures in abstract domains, constrained by other structural correlations between the physical and the abstract domains. This accounts for our capacity to reason about abstract domains such as quantity and purpose.
>
> - The ability to form complex concepts and general categories using image schemas as structuring devices. This allows us to construct complex event structures and taxonomies with superordinate categories.[28]

Thus, the preconceptual level, which is made of basic-level categories and image-schematic structures, and the structuring potential of image schemas, create the basis of conceptual structures. In Lakoff's account, image schemas are crucially important since they play "two roles: They are concepts that have di-

---

[25] *Ibid.* p. 271.

[26] For more details, see *ibid.* pp. 271-278.

[27] *Ibid.* p. 281.

[28] *Ibid.* p. 281.

rectly understood structures of their own, and they are used metaphorically to structure other complex concepts."[29]

Language, thought, and the world are unified due to the switch of perspective. That is, the recognition that "the body is in the mind" relates the extended and the thinking things to each other via motor activity in an obvious way.[30] Accordingly, cognitive capacity has its foundation in a responsive relation to the physical environment. Concepts gain their importance as an effective organizing means, not only in the case of abstract thought, but even in the case of acting upon the world in a less strict sense.

> Only a creature with conceptual abilities can draw consequences from her perceptual processing of visual stimulus...'[T]here is no such thing as deductive inference *between* perceptions'. Upon seeing a brown dog, one can see at once that the animal one faces is a dog and that it is brown. If one perceives a brown animal and one is told that it is a dog then one can certainly come to believe that the brown animal is a dog or that the dog is brown. But on this hybrid epistemic basis, one thinks or believes, but one does not *see* that the dog is brown. One came to know that the dog is brown by seeing it. But one did not come to know that what is brown is a dog by seeing it. Unlike the content of concepts, the content of visual percept is not a matter of inferential role.[31]

Visual perception has some unique characteristics compared to thought: visual perception depends on spatial structures in a strict sense. Another, (but related), difference between imagery and conceptual processing is that thoughts are made of "constituents that can be combined and recombined to generate indefinitely many new structured thoughts."[32]

This characteristic of thought is closely related to language. Lakoff differentiates five idealized cognitive models: image-schematic, propositional, metaphoric, metonymic, and symbolic. We can see the function of image schemas, and metaphoric and metonymic models describe ways of mapping. Lakoff defines the propositional Idealized Cognitive Model (ICM) as "one that does not use imaginative devices, i.e., metaphor, metonymy, or mental imagery." He notes that

---

[29] *Ibid.* p. 283.

[30] Cf. Dewey's distinctions of brain and the rest of the body related to the traditional dualism of body and mind.

[31] Pierre Jacob and Marc Jeannerod, *Ways of Seeing. The Scope and Limits of Visual Cognition* (2003), Oxford: Oxford University Press, 2004, p. 21. Cf. Hume's skepticism. Although recent cognitive psychological research gives place and even some function to Humean skepticism, it at the same time resolves it with the help of an elaborate reconstruction of the cognitive functioning.

[32] *Ibid.* Cf. the objectivist tradition that considers categories as building blocks. That is, we can consider it an argument for the objectivist flavour of the propositional model.

[p]ropositional models have an objectivist flavour, since they contain entities, with their properties and the relations holding among them. ... It seems to me that when we understand our experience by projecting propositional models onto it, we are imposing an objectivist structure on it.[33]

That is, since propositions have their own structure, which is only indirectly related to visual perception and motor activity, they are capable of modifying the world view.[34] Lakoff speaks about symbolic models "[w]hen linguistic elements are associated with conceptual elements in ICMs".[35] The linguistic outlook concludes that "general principles relating semantic to syntactic categories do exist",[36] i.e. basic perceptual experiences create the basis of a semantic scenario and have an influence on semantic-syntactic relations. However far-reaching the structural power of perceptual and motor experience, it seems nonetheless that language is capable of re-constructing basic experiences.[37]

As Lakoff emphasizes, "the locus of metaphor is not in language at all, but in the way we conceptualize one mental domain in terms of another. ... [M]etaphor (that is, cross-domain mapping) is absolutely central to ordinary natural language semantics".[38] To reach abstract reasoning, cognition relies on perceptual and motor experiences; and thanks to cross-domain mapping and the projection of basic-level categories, concepts emerge. But these concepts gain symbolic form, i.e. they are expressed with words, and these words are arranged in accordance with a certain structure.

The term objectification as used by István Hajnal[39] precisely describes what happens in the case of conceptual processing: being able to express something means certain mediation, and the mediator (language in this case) has its own rules of functioning. Accordingly, distortion, at least to some extent, is a necessary concomitant of objectification. And Hajnal uses objectification in a wider sense. He considers technical/technological inventions and institutional arrangements as objectification as well.

Objectification and scaffolding are two sides of the same coin. The term 'scaffolding' emerged in the realm of developmental psychology and pedagogy. Bruner used it to express a kind of interactional support within the framework of

[33] Lakoff, *Women, Fire, and Dangerous Things*, p. 285.

[34] For more on the distorting character of language, see "Old Patterns, New Bewitchments".

[35] Lakoff, *Women, Fire, and Dangerous Things*, p. 289.

[36] *Ibid.*, p. 290.

[37] Cf. the comparison of pictorial and verbal representation in chapters: "The Iconic Turn in Metaphysics" and "*Seinsvergessenheit, Bildsvergessenheit...*".

[38] George Lakoff, "The contemporary theory of metaphor", in Andrew Ortony (ed.), *Metaphor and Thought*, Cambridge Univ. Press, 1993, p. 203.

[39] See the chapter above "Objectification or Spiritual Involvement", esp. pp. 4 f. and 7 f.

adult-child dialogue in the process of learning.[40] Andy Clark expands the meaning of 'scaffolding'. He proposes that

> the old puzzle, the mind-body problem, really involves a hidden third party. It is the mind-body-scaffolding problem. It is the problem of understanding how human thought and reason is born out of looping interactions between material brains, material bodies, and complex cultural and technological environments.[41]

That is, due to the mechanism of scaffolding, the environment, the basic context of everyday life, gains an important role in cognitive processes.[42] Clark refers to several technical inventions that help handicapped people in a variety of ways. Besides these well-designed customized solutions to various impairments, Clark recalls some everyday examples. "Pictures and spoken words, then written words and diagrams, and most recently the full firepower of interchangeable digital media rank high among the tools by which we press maximum problem-solving power from brains like ours."[43]

We know from Donald that the emergence of the external symbolic storage system implied deep cultural and cognitive changes. Let me quote:

> All forms of human representation, from our archaic episodic experiential base, through mimesis and speech, to our most recent visuographic skills, are now refinable and expandable by means of electronic devices. Our minds are thus hybridizations, highly plastic combinations of all previous elements in human cognitive evolution, permuted, combined, and recombined.[44]

---

[40] "Thc concept of scaffolding derives from the work of Jerome Bruner and is, in turn, derived from the Vygotskian theory. Vygotsky proposed a very simple but powerful theory of learning and development. ...It is best stated in his famous statement of the "zone of proximal development": the distance between a child's actual developmental level as determined by independent problem solving and the higher level of potential development as determined through problem solving under adult guidance or in collaboration with more capable peers. This is not just a theory of assessment of development but also a theory of teaching. Bruner developed this into the notion of scaffolding and fading, and David Wood developed this further into his theory of contingent instruction". (Guidelines for learning in mobile environment. MOBIlearn p. 15) For more details see http://www.mobilearn.org/download/results-/public_deliverables/MOBIlearn_D4.1_Final.pdf.

[41] Andy Clark, *Natural-Born Cyborgs. Minds, Technologies, and the Future of Human Intelligence*, Oxford/New York: Oxford University Press, 2003, p. 11.

[42] Cf. "We move, perceive, and think in a fashion that depends on techniques rather than on wired-in arrangements in our nervous system." (Bruner et al, *op. cit.*, p. 56.)

"We depend for survival on the inheritance of acquired characteristics from the culture pool rather than from gene pool. Culture then becomes the chief instrument for guaranteeing survival, with its techniques of transmission being of the highest order of importance." (Bruner et al, *op. cit.*, p. 57.)

[43] Clark, *Natural-Born Cyborgs*, p. 75.

[44] Donald, *Origins of the Modern Mind*, pp. 355 f.

Donald emphasizes the importance of basic representational skills regarding social institutions.[45] However, as we can see above, basic-level perceptions play a highly significant role both in cognitive and motor activity, and as Lakoff notes, "[m]uch of our technology is aimed in the direction of expanding basic-level experience".[46] This provides further crucial support for the interrelatedness of science and folk-theory, and basic and metaphoric context. "It is the technological extension of basic-level perception and manipulation that makes us confident that science provides us with the real knowledge."[47] Clearly, the success of any scientific discovery, and the trickling down into common consideration of any scientific theory, is closely related to technology. That is, our environment in a technological sense determines the framework of perceptible phenomena, and what is accessible via perception is accordingly more convincing and acceptable from a conceptual stance as well.

Conclusion

As we can see, both from the point of view of metaphysics and from recent results in the field of cognitive psychology, context bears significant structuring potential regarding cognitive capabilities. Lakoff's effort to redefine classical categorization has far-reaching consequences: there is a need to reconfigure our vocabulary, or at least to redefine some central concepts.[48] The model that Lakoff provides in opposition to classical categorization is built on cognitive and linguistic research and suggests a much more flexible, although complicated, structure for understanding the process of categorization. Some elements of his argument, such as basic-level categories and their gestalt structure, embodiment, and the special character of propositional ICMs, are crucial in highlighting the role of context both in metaphysics and epistemology. However, the question arises: How do we set boundaries between basic context (i.e. that of motor actions) and metaphoric context (that of conceptual processing)? Is it possible at all?

With certain functions and biological conditions, it makes sense to differentiate the two levels. As Donald put it:

---

[45] "But it is common, especially in large, anonymous crowds of people, to behave in a relatively pure mimetic manner. Thus, in church, in stadiums, in large demonstrations, in civic celebrations, and in various collective rituals humans behave in ways that belie an unreflective understanding of social order, and in ways that are collectively representational, but not linguistically based. In such situations, language is often relegated to a support function, and the entire framework of social exchange is mimetic." (*Ibid.*, p. 369.)

[46] Lakoff, *Women, Fire, and Dangerous Things*, p. 298.

[47] *Ibid.*

[48] See *ibid.* pp. 9, 56, and 292-303.

> Mimesis can be cleanly dissociated from the symbolic and semiotic devices upon which modern culture depends. It serves different functions and is still far more efficient than language in diffusing certain kinds of knowledge; for instance, it is still supreme in the realm of modeling social roles, communicating emotions, and transmitting rudimentary skills. It is also dissociable in terms of its brain representation. Mimetic skill usually survives the disruption of symbolic language.[49]

As we can see there are significant differences between the various kinds of representational and symbolic systems. Thus it is useful to summarize and compare, let's say the verbal and pictorial ways of expression. However, we can realize that some key characteristics can come to light only with the involvement of basic context.

Fusing together metaphysics and epistemology requires handling them as a whole. Objectifications, according to Hajnal, are the products of conscious and abstract thinking, and they take part in further abstract considerations as the term 'scaffolding' suggests.

Bruner's considerations excellently perform the mutual effect of the two kinds of context. As he wrote, "because so much of the 'meaning' of what is being learned is intrinsic in the context in which the learning occurs, there is very little need for verbal formulation". Bruner goes on to quote from Fortes: "It would seem that Tale children rarely have to ask 'why' in regard to the people and things of their normal environment because so much of their learning occurs in real situations".[50] We can say that the more complex a society is, the stronger the need for "*telling* out of context rather than *showing* in context."[51] This educational perspective also shows that human activity is deeply intertwined /tangled with its complex environment. To divide the two kinds of context seems a hindrance to understanding cognitive, epistemic, or even social mechanisms.

Wittgenstein's self-mocking reminiscences in the *Philosophical Investigations* form as it were a prism in which the foregoing is refracted.

> Thought is surrounded by a halo. Its essence, logic, presents an order, in fact the a priori order of the world: that is, the order of *possibilities*, which must be common to both world and thought. But this order, it seems, must be utterly simple. It is prior to all experience, must run through all experience; no empirical cloudiness or uncertainty can be allowed to affect it. – It must rather be of the purest crystal. But this crystal does not appear as an abstraction; but as something concrete, indeed, as the most concrete, as it were the hardest thing there.[52]

---

49 Donald, *Origins of the Modern Mind*, p. 198.

50 Bruner et al., *op. cit.*, p. 61.

51 *Ibid.* p. 62.

52 Wittgenstein, *Philosophical Investigations*, § 97.

Converging Theories: The Age of Secondary Literacy*

The notion of secondary literacy can best be approached by pointing to the fact that although alphabetic rational reasoning is dominant in scientific and even everyday thinking, communications technology is capable of mediating experiences as if they were direct, i.e. without verbal mediation. (I am using the term "secondary literacy" in a sense different from that introduced by, on the one hand, Walter J. Ong and, on the other, Wolfgang Coy.[1]) The pressure to formulate experiences verbally is decreasing; the birth of a new rationality is imminent. In the age of literacy, where alphabetical writing is the dominant medium (for expressing and preserving knowledge and ideas), there is, due to the special character of language and that of writing itself, a certain detectable convergence of methodologies and the thought processes which are tailored in accordance with the ruling principles of mathematics and logic. However, this convergence is accompanied by opposing traditions and unsolvable paradoxes. In the age of secondary literacy, traditionally opposed schools of philosophy, as well as new claims implied by recent scientific and philosophical inquiry, have at some points been converging. New mobile devices that condense all the previous inventions of communication technology play an important role in this process because they are capable of becoming the primary mediator of experience. This capability opens a path towards developing a common conceptual background that can help theories converge. In the present paper, I will first outline the main characteristics of the literate mind; then I will attempt to display some converging tendencies, focusing on philosophy and certain recent scientific self-analyses. Finally, before concluding my argument, I will, briefly, draw attention to some new issues pertaining to the importance of perceptual experience with regard to abstract reasoning.

---

* This is a revised version of a talk given in September 2007 in Budapest at a conference *Towards a Philosophy of Telecommunications Convergence*.

[1] With regard to W. J. Ong see "Schmandt-Besserat on Writing and Consciousness – Ong Sessions at MLA 2006 and CCCC 2007. Ong on Secondary Orality and Secondary Literacy", http://ongnotes.slu.edu/?p=190; and "Secondary Orality and Secondary Visualism", http://-libraries.slu.edu/sc/ong/digital/texts/lectures/lecture1.pdf. As to Wolfgang Coy, see his "Text and Voice – The Changing Role of Computing in Communication", in Kristóf Nyíri (ed.), *Mobile Communication: Essays on Cognition and Community*, Vienna: Passagen Verlag, 2003, pp. 93 f.

Converging Methodologies

The emergence of language in the course of our phylogenesis has far-reaching effects both from a cognitive and a social point of view. In accordance with the *social brain* hypothesis,[2] language as a new representational system was able to create a cohesive background that furnishes the main rules for everyday activity, and criteria for group identity. This was possible because, via language, thoughts become objectified: sharing and handling these thoughts is made much more effective with the help of fixed symbols, specifically with the help of words. Also, since words tag categories, in a way the processing of perceptions becomes quicker. Being able to represent experiences or thoughts via language means, at the same time, being able to reconstruct them in accordance with the limits of verbal expression. This implies that experiences/thoughts must be linearly ordered, just as words are sequentially arranged. The other decisive feature of language is that it mediates the demands of the speaker. In accordance with the social brain hypothesis, we talk to others primarily in order to cooperate/coexist with them, and thus to influence their actions according to our demands. Turning to ontogenesis, the verbal perspective is strictly bound to the given situation: learning how to use words in accordance with our purposes, and understanding the words of others, is possible through a practice which is embedded into the given situation.

Alphabetical writing, which is a highly effective and successful *external symbolic storage system*,[3] makes the above-mentioned characteristics of language more radical, since written words are necessarily divested of much of their original context. The significant details of the situation have to be verbalized and the linear structure is even stricter because there is no possibility of referring to previous or coming details with the help of gestures or facial expressions. Moreover, since written texts are created to preserve knowledge, they must remain unambiguous across radically different circumstances.

After the invention of alphabetical writing, in the early stages of its use, some momentous ideas emerged in which reflections on the advantages and limits of writing were implicit. Heraclitus complained about the perceived inability to express timeless and general truths. Let us point out that Heraclitus' problem had to do with a lack of appropriate concepts. In the form of *logos*, he presupposes a higher and eternal intelligence with whom we are theoretically capable to cooperate. Plato talks about the thing-like character of writing, which fallaciously suggests that knowledge is possible outside the human soul. These

[2] See Robin Dunbar, "Are There Cognitive Constraints on an E-World?", in Kristóf Nyíri (ed.), *Mobile Communication*, p. 58.

[3] See Merlin Donald, *Origins of the Modern Mind: Three Stages in the Evolution of Culture and Cognition*, Cambridge, MA: Harvard University Press, 1991, pp. 308–325.

early deliberations reveal some important characteristics of the literate mind. Heraclitus' demand for highly abstract concepts, the presupposition of a higher intelligence which can be mediated exclusively via abstract concepts, as well as Plato's claim that knowledge is bound to the soul and is opposed to thing-like entities, create the basic presuppositions of western metaphysics. There gradually evolved a rather complex and abstract framework of concepts, and systematic methodological principles emerged to serve the recording and interpretation of verbalized thoughts. This systematic analyzing attitude and the expectation of clear-cut categorization complemented by the ruling principles of logic and mathematics create a special one-sidedness in philosophical thinking. Reason was comprehended as being separate from practice or any engagement with the external world. This separation of practice and reasoning, and that of the body and mind, leads to paradoxes and, of course, diverging attempts to solve them. (Consider, for example, the tradition of empiricism vs. rationalism, or the later tradition of the logical positivist criticism of metaphysics.)

The philosophical analysis of language implies permanent *mental discomfort*[4] because of the lack of a satisfactorily transparent and clear-cut way of expression. Wittgenstein's effort to grasp how language works reveals the need for a shift which embeds language into everyday practice. Consequently, several important features of language come to light.

Reflecting upon how language works, and the characteristics of language giving rise to specific philosophical difficulties, Wittgenstein concludes that, misled by grammar, we are often inclined to use substantives as if they denoted things.[5] Just as in the case of the "mind",

> [t]here are the sounds of the words, and all sorts of bodily sensations connected with gesture and intonation. Where we are liable to go wrong is in supposing that sensations connected with words are somehow 'in the mind'. The phrase 'in the mind' has caused more confusion than almost any other in philosophy.[6]

---

[4] This was a phrase used by Wittgenstein: "The problem may seem simple, but its extreme difficulty is due to the fascination which the analogy between two similar structures in our language can exert on us." (Ludwig Wittgenstein, *The Blue and Brown Books*, Oxford: Basil Blackwell, 1958, p. 26.) See also: "we find that there is puzzlement and mental discomfort, not only when our curiosity about certain facts is not satisfied or when we can't find a law of nature fitting in with all our experience, but also when a notation dissatisfies us – perhaps because of various associations which it calls up." (*Ibid.*, p. 59.)

[5] "One of the chief troubles is that we take a substantive to correspond to a thing. Ordinary grammar does not forbid our using a substantive as though it stood for a physical body." (*Wittgenstein's Lectures, Cambridge, 1932–1935*, ed. by Alice Ambrose, Oxford: Basil Blackwell, 1979, pp. 31 f.)

[6] *Wittgenstein's Lectures,* p. 114.

The other key characteristic of language is closely related to its usage. As Wittgenstein puts it, "our language is tempting us to draw some misleading analogy. This should remind us of the case when the popular scientist appeared to have shown us that the floor which we stand on is not really solid because it is made up of electrons".[7] There are different language games, and the necessity of switching between them is not always recognized. This observation might suggest that being able to communicate multimodally, i.e. not only via language, would make these switches much easier to handle. Multimodality makes the interpretation of a given perception less ambiguous. Consider Wittgenstein's example: observing the floor and observing the elementary particles making up the same (with the help of previous experiences and some knowledge of physics) renders the difference between the two kinds of solidity obvious, i.e. it clearly shows the misleading nature of analogies engendered by language.

Towards a New Rationality

Secondary literacy as the age of a heightened rationality of the literate mind complemented by multimodality strengthens the tendencies that were emerging since the late 19th and early 20th centuries. The birth of photography and, later, that of film, the invention of the telegraph, the telephone, and the gramophone, just to mention a few new devices of the time, had considerable impact on what were, until then, apparently immutable concepts and accustomed dualisms.

Being able to record not only propositions, but images and situations (thanks to photography), and being able to mediate live voices (due to sound recording and transmitting technology), enables us to return to the vivid mundane world of everyday activity as opposed to the mute and distancing world of texts. Casting a glance upon the history of philosophy, we can now better understand what it means when the emerging philosophy of pragmatism suggests taking into consideration the practical issues of life as opposed to mere theoretical reasoning. Dewey formulated the demand for a new theory of knowledge "which sees in knowledge the method by which one experience is made available in giving direction and meaning to another".[8] Dewey notes that traditional dualisms imply a certain one-sidedness in practice, e.g. the traditional opposition of mind and body in practice means a focus on only one of the pair at a time.[9] The importance of the environment and its close relation with the human mind is another crucial insight of the time. Think of Henri Bergson's effort

[7] Ludwig Wittgenstein, *The Blue and Brown Books*, Oxford: Basil Blackwell, 1958, p. 48.

[8] John Dewey, *Democracy and Education* (1916). *The Middle Works of John Dewey, 1899–1924*, vol. 9, Carbondale and Edwardsville: Southern Illinois University Press, 1985, pp. 354 f.

[9] *Ibid.*, p. 346.

to resolve the above-mentioned dualism by calling attention to the role of perception and action, which places human beings within the framework of everyday activity rather than within that of reasoning.

Decades later, this notion of embeddeness played a central role in Heidegger's philosophy. Being-in-the-world means: we are engaged in a certain activity and are committed to this acting upon the world. The emphasis on *thrownness* into the world furnishes the basis for a criticism of mainstream artificial intelligence research with roots in logical positivism. AI research is based on the conviction that human reasoning is a symbol-manipulating process: each element of a thought process can be formalized in accordance with the rules of logic. The first steps were based on an attempt to define the nature of knowledge with the same formal rigour that was then popular in the world of mathematics. This conviction is possible because the focus is on theoretical knowledge that takes form in a logical order of propositions. Terry Winograd and Fernando Flores stress the importance of

> the shift from an individual-centered conception of understanding to one that is socially based. Knowledge and understanding (in both the cognitive and linguistic senses) do not result from formal operations on mental representations of an objectively existing world. Rather they arise from the individual's committed participation in mutually oriented patterns of behavior that are embedded in a socially shared background of concerns, actions and beliefs. This shift from an individual to a social perspective – from mental representation to patterned interaction – permits language and cognition to merge.[10]

That is, the authors try to assimilate some of the main concerns of Heidegger, a prominent figure representing the alternative conviction.

Considering the different views on perception in, for example, the field of epistemology, a kind of convergence emerges. On the basis of the traditional mind–body dualism and with the dominance of theoretical knowledge, diverging solutions and solid anomalies characterize this field (consider sense-data theory and the problem of mind-independent objects, or direct realism and/or disjunctivism and the burden of skepticism). Convergence emerges due to the change of focus effected by the representational theory of the visual mind (RTVM), according to which there are two kinds of visual representations:

> visual percepts and visuomotor representations. The former serves as input to higher human cognitive processes, including memory, categorization, conceptual thought and reasoning. The latter is at the service of human action. From the standpoint of our

[10] Terry Winograd – Fernando Flores, *Understanding Computers and Cognition. A New Foundation for Design*, New York: Addison-Wesley publishing Company, Inc., 1986, p. 78. See also my review of this book: Zsuzsanna Kondor, "Dasein és design", in Világosság, 1989/11, pp. 879 f.

version of the 'two visual systems' hypothesis, vision serves two masters: thinking about, and acting upon, the world.[11]

The change of focus means the inclusion of action.

> We like RTVM precisely because we think that it avoids the pitfalls of both sense-datum theory and disjunctivism. RTVM is often called 'intentionalism' because it makes the basic claim that the content of a visual experience crucially depends upon the 'intentional' properties that the experience represents objects as having. According to the teleosemantic version of visual intentionalism ... the human visual system has been selected by evolution for carrying information about (or for processing) a particular class of properties instantiated in the environment of early humans. According to visual intentionalism, what matters to a visual experience – what makes it the experience that it is – are the properties that are represented in the experience, not the particular objects that happen to exemplify the properties.[12]

Directing attention to primordial functions sheds light on the misleading character of the traditional framework. The role of environment is obvious if we consider that acting upon the world is primary compared to abstract reasoning. And the same attention reveals that it is not separate objects with solid contours, but the situation and certain details that have importance.[13]

Converging Experiences

Higher cognitive functions, such as reasoning, need categories, sometimes highly abstract categories. According to conceptual metaphor theory, categories, even highly abstract ones, are related to mundane everyday experiences and activities. As Lakoff put it, "conceptual structure is meaningful because it is embodied, that is, it arises from, and is tied to, our preconceptual bodily experiences. In short, conceptual structure exists and is understood because preconceptual structures exist and are understood."[14]

According to Lakoff, the preconceptual level is determined by kinesthetic image-schematic structures[15] and basic-level categories. The latter do not over-

---

[11] Pierre Jacob – Marc Jeannerod, *Ways of Seeing: The Scope and Limits of Visual Cognition*, New York: Oxford University Press, 2003, rev. ed. 2004, p. 45.

[12] *Ibid.*, p. 16.

[13] Compare Lakoff's concept of categorization. He suggests that categories are not comprehensible as sets. Rather there are "fuzzy boundaries", prototypes, different degrees of membership, contrasting elements, close relations to the experiential world, exceptions, etc. See George Lakoff, *Women, Fire, and Dangerous Things: What Categories Reveal about the Mind*, Chicago: The University of Chicago Press, 1987.

[14] *Ibid.*, p. 267.

[15] Kinesthetic image schemas originate from everyday bodily experiences and mean certain, relatively simple spatial relations such as: containers, paths (movement in space with a

load memory capacity and are closely bound to motor activity. (Consider, for example, the basic-level category of "chair". It has many subordinate categories, i.e. there are a great many kinds of chairs, which might imply a considerable load on memory. And to its superordinate category, "furniture", there is no chance of attaching one special motor action.)[16] In Lakoff's account, image schemas are crucially important since they play "two roles: They are concepts that have directly understood structures of their own, and they are used metaphorically to structure other complex concepts."[17]

As we can see, perception plays a crucial role both in acting upon and in gathering information for thinking about the world. Issues related to perception are also vitally important in the development of science.[18] The potential of perception to persuade or convince is considerable, both with regard to the science/folk-theory relation and the horizon/limit of perceptible phenomena. That is, the extension of perception can open up new perspectives. Of course, specialization means that different kinds of methods and technical accessories emerge, but converging questions arise due to possibly common perceptual experience.

Think of the question of reality in the mobile age. Is there any reason to distinguish virtual and real in the case of a location-sensitive and at the same time location-independent device? What should be considered real: the given circumstances and the concomitant tasks, or those engendered by the mediated-from-afar via mobile devices? Is there any relevance to the distinction between real and virtual when considering function and effect?

Not only the social sciences and the humanities face these questions. A similar question emerges in the field of quantum physics. The intuitive idea of so-called *local realism* (an idea of course very much bound up with Einstein's discoveries) suggests "that a particle cannot be instantly influenced by a distant event, and that its properties exist independently of any measurements".[19]

---

starting and ending point including direction), links (pertaining to emotional security and its sources), forces, balance, different kinds of orientation (up-down, front-back, part-whole, center-periphery, etc.) For more details, see *ibid.*, pp. 271–278.

[16] For a short introductory summary of basic-level categories, see Lakoff, *Women, Fire, and Dangerous Things*, p. 46.

[17] *Ibid.*, p. 283.

[18] "The whole underlying problem, ultimately, is that we lack experimental observations in the region where quantum and gravitational effects both matter" – says physicist David Deutsch (quoted by Michael Brooks, "Reality Check", in *New Scientist*, 23 June 2007, p. 33). That is, even in the highly mathematized realm of physics, there is a need for perceptual support.

[19] *Ibid.*, p. 31.

Although the implications of quantum physics[20] suggest that "[t]he world could not be local and real", there arises the question of whether it is either local or real. The various experiments supporting quantum theory led to the conviction that it is possible that "there is nothing inherently real about the properties of an object that we measure. In other words, measuring those properties is what brings them into existence... So does the universe exist independently of measurements?"[21]

We can see that similar difficulties emerge in the field of epistemology with regard to perceptions and higher cognitive functions. This shift in the focus of philosophy overwrites some crucial notions, and a similar demand has been formulated by Caslav Brukner with regard to physics: "We need to rethink and radically revise our basic physical concepts before we make the next big breakthrough in physics."[22] And this seems to be obvious if we think of the conceptual framework in which we try to grasp the consequences of quantum physics. The concepts of this framework are tailored according to physical experiences which are interpreted in accordance with the literate mind that is striving for clear-cut categories.[23]

Conclusion

When everyday experience can involve confusion as to reality and locality, while there is also a tension between locality and reality in the field of physics, our traditional conceptual framework itself can be queried. The philosophical rediscovery of the importance of perception and action, and their relation to abstract reasoning, suggests a reconfiguration of some crucial concepts. Mobile technology catalyzes this process, since in the broad scope of everyday experiences perceptions suggest that some familiar constellations have been changing. Communication complemented with multimodal accessories utilizes the obvious power of visually-gained information (although we are often unaware of this), and accordingly makes it easy to mediate experiences non-verbally. Finally, since mobile devices are capable of supplanting most previous communication-technology inventions, they can furnish a common framework of experience.

---

[20] "In our experience, objects have definite location in space and a limited range of influence. According to quantum theory, however, a pair of particles would be able to share information about each other's quantum states – and sometimes influence them – even where the distance and timing involved meant that no signal could have passed between them." (*Ibid.*, p. 31.)

[21] *Ibid.*, pp. 32 f.

[22] *Ibid.*, p. 32.

[23] See the classical categorization recapitulated by Lakoff, *Women, Fire, and Dangerous Things*, p. 152.

INDEX

Peter Baltes

Critical Theory of Good Living

Frankfurt am Main, Berlin, Bern, Bruxelles, New York, Oxford, Wien, 2005.
170 pp., num. fig.
ISBN 978-3-631-54320-7 · pb. € 41.10*

To understand our life, to order its elements, to add new ones, and to orient it toward the general goal of a good life for all: this is the task of this Critical Theory of Good Living. Learning from *Newton* that the world is a system of forces, and from *Hegel* that all forces are subject to reciprocal effects, it was *Parsons* who could order the world according to the significance of every force in our life. *Darwin* discovered the self interest as the intrinsic motive force of our behaviour, dialectically completing *Kant,* who had proposed equal rights as the absolute moral law for all human action. From the integration of these fundamental tenets of thought into one logical system emerges this new and comprehensive philosophy, which not only includes a superior life technique but also the principles of a justified moral existence.

Contents: Mutual action · The life structure · The life concept · The GACORE-Model · Self-advice · Argumentative communication · Production and consumption · The model of the *Critical Theory of Good Living*

Kolibri 360 GmbH
Ferdinand-Jühlke-Straße 7
99095 Erfurt, Deutschland
produktsicherheit@kolibri360.de

Druck:
CPI Druckdienstleistungen GmbH
im Auftrag der
Kolibri 360 GmbH
Ferdinand-Jühlke-Str. 7
99095 Erfurt